A Memory of Sky

A PILOT'S VIEW OF CANADA'S CENTURY OF FLIGHT

JIM SHILLIDAY

GREAT PLAINS
PUBLICATIONS

Copyright © 2009 Jim Shilliday

Great Plains Publications
345-955 Portage Avenue
Winnipeg, MB R3G 0P9
www.greatplains.mb.ca

All rights reserved. No part of this publication may be reproduced or transmitted in any form or in any means, or stored in a database and retrieval system, without the prior written permission of Great Plains Publications, or, in the case of photocopying or other reprographic copying, a license from Access Copyright (Canadian Copyright Licensing Agency), 1 Yonge Street, Suite 1900, Toronto, Ontario, Canada, M5E 1E5.

Great Plains Publications gratefully acknowledges the financial support provided for its publishing program by the Government of Canada through the Book Publishing Industry Development Program (BPIDP); the Canada Council for the Arts; as well as the Manitoba Department of Culture, Heritage and Tourism; and the Manitoba Arts Council.

Design & Typography by Relish Design Studio Inc.
Printed in Canada by Friesens

Library and Archives Canada Cataloguing in Publication

Shilliday, Jim, 1928-
A memory of sky : a pilot's view of Canada's century of flight / Jim Shilliday.

ISBN 978-1-894283-95-3

1. Aeronautics--Canada--History. I. Title.

| TL523.S45 2009 | 629.130971 | C2009-903322-4 |

Always for Beth, who contributes so much, and
can't imagine a life not surrounded by books;

and

To Gregg and Ingeborg; Caron and Mervin; Charles
and Paula; Thomas and Kath; and Catherine—for listening to my "hero" stories;

and

To my big brother, Flight-Sergeant Robert Charles Shilliday, R263945, RCAF, No. 153 Squadron,
RAF Bomber Command, Lancaster mid-upper gunner, killed Jan. 16, 1945,
age 19, during night operations over Zeitz, Germany;

also

To all those who fly, but particularly to the hundreds of former Sword drivers,
dead and alive, who for a brief glorious time were part of the grandest
Canadian fighter group ever, and were the best in the world.

FLIGHT RECORD

	Type	No.	1st Pilot	or Passenger	(Including
7	Spitfire	R6919	self	—	Patrol
7		R6919			M.H. to Co
7		R6919			Coltishall
8		R6958			Test a/c
11		R6919			To Coltish
13		R6919			To Duxfo
15		R6919			Patrol.
15		R6919			Patrol
16		R6919			

AIRCRAFT FLOWN

CERTIFICATE NUMBER	ENGINE	FROM	TO
17 ARMY	ALLISON V-1710-73	HENGYANG	NANCH
18 "	R6923 "	KWEILIN	HENGY
19 "	R6933 "		
22 "	R6919 "	HENGYANG	SUICH

Yet for a space we held in our
Morning's hand
the welling and wildness of Canada,
the fling of a nation
We who have ridden the wings of our
people's cunning
and lived in a star at peace
among stars
return to our ferment of earth with a
memory of sky.

Excerpted from the long poem, *North Star West* (1951)
by Earle Birney

CONTENTS

FOREWORD 9

PRE-TAKE OFF CHECK • Blue Skies Beckon 11

CHAPTER ONE • First Flight, First Pilot 15

CHAPTER TWO • The First World War 27

CHAPTER THREE • Civil Aviation 49

CHAPTER FOUR • Coming of Age: Taming the Atlantic 65

CHAPTER FIVE • The Second World War 81

CHAPTER SIX • The Cold War 97

CHAPTER SEVEN • Time of Transition 143

CHAPTER EIGHT • Aerospace: A Powerful Canadian Symbol 153

POST-LANDING CHECK • Vintage Wings and Hawk One 165

FOREWORD

JIM SHILLIDAY'S DEFINITIVE book — *A Memory of Sky* – prepared in such magnificent detail, is a must read for all who have enjoyed the thrill of flight, as well as for other Canadians who will come to appreciate through this book the great role that aviation has played and continues to play in the development of our country and its importance in our history.

He has thoroughly documented, in a very appealing and readable style and with great feeling, the evolution of this rich Canadian heritage since its earliest beginnings at Baddeck, Nova Scotia, in 1909, up to the very vibrant and active current day situation.

Like Jim, I started my flying with the Royal Canadian Air Force as a young nineteen-year-old in the early 1950s. Also like Jim, I trained on the great Harvard — often referred to as the Yellow Peril — and then had the very early grand experiences of flying some of Canada's jet fighter and training aircraft, and in particular that wonderful Canadian-built, Orenda-powered, Canadair F-86 Sabre, both the Mark 5 and 6 versions.

My flying experience on the Sabre spanned some eight years, including almost six years in Germany during the tense 1950s and early 1960s period of the Cold war, when we were prepared to do battle as part of NATO against the Russian Bear.

My six years flying the Sabre in Europe comprised over three and one-half years on 422 (Fighter) Squadron at 4 (Fighter) Wing Baden Soellingen, followed by a further gratifying period of just over two years as a Tactical Advisor to 73 Wing of the newly constituted West German Air Force.

Like Jim, these early experiences as a young man, flying the top of the line Sabre, in a very challenging environment, instilled a lasting impression and passion in me about the thrill of flight and the on-going and highly important mission of Canada's Air Force.

Here I am fifty-seven years since my first training mission in the Harvard at RCAF Station Claresholm, Alberta, and I am just as keen as I was as a young man, still stimulated by my many earlier experiences. I still fly regularly (although no longer with swept wings and not at a rate approaching the speed of sound) and I maintain my commercial pilot license and instrument rating.

I am also very enthused to have become involved as a technical advisor to the outstanding Hawk 1 project — that wonderfully restored Sabre 5 owned by Vintage Wings of Gatineau, Quebec — an integral part of the celebration of the Centennial of Flight in Canada.

Jim Shilliday's book is an inspiring portrayal of the wonderful history of aviation in Canada over the past one hundred years. I commend *A Memory of Sky* to all. It is clearly an important literary work and a great addition to the national Centenary of Flight in Canada.

Paul A. Hayes, OMM, CD
Brigadier-General (Retired) Royal Canadian Air Force

Paul Hayes (right), in 1961 Tactical Advisor to JG 73 (Jagdgeschwader, or Fighter Wing 73) discussing flying operations with German Air Force pilot Leutnant Horst Fetzer. The new West German Luftwaffe set up a Waffenschule (flying school) at the northern West Germany city of Oldenburg, in 1958. Canada provided Germany with Sabres, T-33s and Harvards, plus leadership, technical, administrative and aircrew support to get the school going.

PRE-TAKE OFF CHECK

The author prior to first solo in the Sabre jet. Face visors weren't yet available, and goggles of an earlier era were the order of the day.

BLUE SKIES
BECKON

THE GREEKS MYTHOLOGIZED IT. Leonardo da Vinci hinted at it, and Jules Verne fantasized about it. But true flight—achieving a Third Dimension of existence on the planet—was a defining moment in the world of the early 1900s.

The flying mystique—"slipping the surly bonds"—captivated Canadians from the beginning. Canada's century in the air has been a major chapter in the history of flying. With its vast and formerly unreachable territories, and consequent winged development, this nation was an original testing ground for all aspects of aviation. The world's flying profession—both civil and military—owes much to the early Canadian aviation pioneers. It owes less to those federal authorities who, from the beginning, often showed disdain and faint-heartedness.

Since those magical days of the wire-and-fabric flying machines when man first climbed into the clouds, Canada has helped show the way. This immense country—flat and undulating, with forests, coastlines and mountain chains—was a natural place for aviation to thrive. No nation in the world owes more to the airplane, which opened up the vastness and provided a significant lifeline between all coasts, and into the massive northern areas. Canada's location meant it was the front-line in early attempts to fly the Atlantic and transform an adventure into a world-shrinking business. Dozens of brave men and women took off from our eastern coast to fly into fame—or death—and others landed on, or bumped into, her shores after flying the ocean from the other direction.

As we will see, aviation began in the Maritimes. Then, for a time, the West was the heart of aviation in Canada. The nation's first airline

JIM SHILLIDAY

was founded in Winnipeg. Prairie pioneer pilots developed bush flying to an art, and inaugurated the first heavy freighting by air, particularly into the North; they were the intrepid first to wing across the Arctic Circle and, like all great explorers, "walk off the map" into the unknown territories marked on rudimentary charts.

Canada's experience encompasses the first faltering attempts to free man from the ground; the barnstormers; the record-breakers; the dollar-a-minute flights; the war years. It takes in the progress from simple glue and canvas to rivet-popping speeds above mach one with the Canadian Orenda jet engine, astronauts, aerospace instruments and the robotic space-and-satellite device, Canadarm. (It is astounding the difference Canadian aeronautical engineers made to the whole U.S. NASA space program and moon landings). Also outstanding is the emergence of Bombardier as the world's third biggest aircraft producer.

It is appropriate that the first Canadian powered flight took place in the Maritimes. So many early aviation events of historical significance occurred in the Atlantic Provinces, and still do.

The specific event that the 2009 centenary of flight observed was the first powered flight in Canada of the *Silver Dart* from the frozen Bras d'Or Lake at Baddeck, Nova Scotia, on February 23, 1909. Douglas McCurdy piloted his fragile plane in what was a first in the British Empire, and a product of the Aerial Experiment Association established by Alexander Graham Bell. Not much had happened since the Wright Brothers' historic 1903 flight, and the Silver Dart's success excited many Canadians.

In his history of flying in Canada, High Flight, Jonathan F. Vance wrote that Canada retains a deep, almost elemental, connection to flying. "How else to explain the storms of anger whipped up by an announcement that the much-loved Snowbirds aerial display team might fall victim to government budget cuts?…And the Avro Arrow? If it had been a revolutionary battle tank or a state-of-the-art submarine, it would long ago have been forgotten. But because it was an airplane, it lives on in our national psyche."

Many Canadians are historical nationalists. They love their country and respect what the nation has accomplished in the past. Traditionally, Canada has been a leader and an innovator in aviation (bush pilots; aerial fire suppression; aerial photography and mineral searches; jet plane designs), its people showing a natural aptitude. But politicians often have not had the nerve or imagination to take advantage of this skill.

MEMBERS OF THE MANITOBA CHAPTER of the Canadian Aviation Historical Society (CAHS) were the genesis of this book, along with the observance of a century of flight in Canada. They asked me, as a former "Sword" driver, to tell them about my experiences while flying with NATO out of England, West Germany and France at the height of the Cold War. What struck me was the intense interest of these aviation enthusiasts consisting of glider pilots, ex-riggers and fitters of the old Royal Canadian Air Force, department managers in small airlines, and those from that cross-country legion of younger people who simply love to know and talk about Canadian aviation, particularly of what it used to be.

After flying the Mark 2, Mark 5 and Mark 6 Sabres during a five-year short-service commission in the mid-1950s, my only flying since has been as a passenger. So my address would not be from an aviation expert, I told the historical society members, but from one who considered his time flying military jets one of the highlights of his life. Some of this book is about my Cold War experiences. For the general Canadian reader, not enough has been written about the period. My memories of that time are strong and good. I hope today's military flyers will have the same good feelings when "now" becomes "then."

There are many excellent sources that I have drawn upon. Larry Milberry's magnificent volumes published by his CANAV Books, Toronto, are essential reading. The CAHS journal, and Airforce, the magazine of the Air Force Association of Canada, edited by Vic Johnson, also contribute firmly to the maintenance of this country's aviation heritage. A relatively new addition to Canada's collection of aviation heritage supporters, but packing a supersonic rate of climb, is Michel Potter's Vintage Wings of Canada Foundation, whose fine pilots, writers and photographers constitute a high-octane, "live museum" of immense value to Canadians.

I should make clear that this is not an in-depth, chronological presentation sweeping up all the aviation facts contained in the spread of one century. Rather, it is a narrative history, a cornucopia of events, national and personal, selected as defining or exciting, all flowing onto a basic presentation of civil and military flying. The passion, the rhapsodic experience of flying written about here applies, of course, to piloting rather than the cramped, bored experience of the airline traveller, whose main concern is, "When will we get there so I can get out of this thing?"

One goal of this work is to throw more light on the existence of Canada's best "peace-time" air force of the century, the RCAF's No.1 Air Division, in Europe, that was an operational battle-ready air force with NATO during the tense 1950s period of the Cold War. The pilots were highly trained to be "The Few," if the Russian bear's claws were unsheathed. Yet, historian J.L. Granatstein dismissed this proud achievement with the yawn that, in the 1950s and 1960s, the air force excelled "in long-range transport" and was "eminently suitable for peacekeeping roles." One military historian opined that a national identity had come of two world wars, a war in Korea and "in a host of minor peacekeeping excursions." Where did that leave Canada's NATO air force of the 1950s, the biggest air force in the country's history but for the Second World War and, without doubt, for a short time sending up the best fighter squadrons anywhere?

Some of the excellent men who flew Sabres when I did had skills and flying experience that I know would make better reading than mine. The history of Canadian aviation owes nothing to this writer; but I owe a lot to the grand business of piloting, because it coloured my early imagination, seduced me into devoting several years of my life to flying jet fighters, and set me up for a lifetime of interest in, and memories of, flying which in varying small ways I have passed on to others. This book is an example. •

Douglas McCurdy poses at controls of a Curtiss biplane in 1911. *Photo: James collection, City of Toronto Archives*

CHAPTER ONE

FIRST FLIGHT, FIRST PILOT

I WALKED UP THE WINDING, tree-lined roadway to Beinn Bhreagh—overlooking Bras D'Or Lake and then the Atlantic—but was stopped from entering the property by a chain with a polite message attached, asking visitors to please respect that this still was private property. But that didn't matter: all the way up, and back, my mind was in another era, time-travelling this path upon which such mental and physical adventurers as McCurdy, Curtiss, Bell and their friends had trod, and marvelling at what a great contribution they had made to Canada's aviation history.

The flying aesthetic manifests itself in uncountable ways. It can be beauty in the eye, and it can be beauty of the mind. Once, when I was a military pilot, I was left with an unforgettable impression after taking off in the dark, then climbing to an evening sunrise as my Sabre's increasing altitude brought the horizon down so that the orb's rays once again lit the immediate atmosphere of my cockpit. The sensation came on suddenly, and it was a bewildering, other-worldly experience – while the Sabre throbbed westward on its night-time climb, the dead sun had been reborn. The burning curvature ahead stimulated my eyes and rendered the aircraft motionless. Behind, and over my head, was a carapace of black that melted to honey towards the unset sun. I was Judgement sitting in an armchair!

Nature played the dominant role in that *tableau vivant* of flight, but nature often is just a setting for the breath-catching delight of man's creative idea—the airplane. Such was the case in 1909 at Baddeck, on the ice of Nova Scotia's Bras d'Or Lake.

With Alexander Graham Bell in the sleigh, and Douglas McCurdy at the controls, the Silver Dart is pulled over the ice at Baddeck for the historic 1909 flight. *Photo: Courtesy Airforce Magazine*

FIRST PILOT

TWENTY-TWO-YEAR-OLD Douglas McCurdy and his flimsy "aerodrome" wavered into the salt-tinged Cape Breton Highlands air on February 23, 1909, and the mantle of First Pilot settled round his shoulders forever. A century later, his feat would be honoured for its significance, and also for its bravery, controlled fear, and skilled determination. All recognizable Canadian traits of the time, for his kind of people were brought up dedicated to the task of making a good person, a good life, and building a strong nation.

His flight—he called it "droming"—was extraordinary because few men in the world had done it, and no one before him had done it in Canada—or in Scotland, Ireland, India, South Africa, Australia or any other pink spot on the world map denoting membership in the British Empire. In the slow-moving times in which the first powered flight in a heavier-than-air machine in the Dominion was made, the flight was daring and speed and accomplishment hard to imagine. It was a great show, and McCurdy said the sensation was like having three or four shots of whisky. He wanted to do it several more times.

The whole countryside had turned out in the brisk winter air to see the attempt. By now, they had

read about the aerial crackpots and cranks, the ludicrous "wing flapping" of some experimental flying attempts, and they had heard that two brothers in the States had "flown" their Wright Flyer for a few seconds, and a few feet, but few people had actually seen anyone fly. One American newspaper headline had scowled of the Wright brothers: "Are They Fliers, or Liars?" So what would those gathered at Baddeck see? "Many of the old-timers came fully prepared to witness a failure, perhaps a tragic one. They had heard of the McCurdy group's flights in Hammondsport, New York, but then the United States was a long way off to them. And even in those times, Maritimers had a quiet distrust of American ballyhoo. School was let out for the occasion, and most of the pupils had the foresight to come on skates," wrote H. Gordon Green, in *The Silver Dart*. Folks had begun gathering at the lake when word spread around the highland community that McCurdy would be performing a flying "experiment" that afternoon.

A local man, John MacDermid brought his one-horse sleigh to pull *Silver Dart* along the ice, assisted by an eight-man crew pushing at the wings and tail. McCurdy had donned a stocking cap and walked alongside. The plane, with its 42-feet wingspan, looked light and airy on its smallish motorcycle wheels (runners were tried later). McCurdy, the aircraft's designer, checked the engine and controls, impatient to get going. Then Alexander Graham Bell himself drove onto the ice in an impressive bright-red cutter, and sporting a mountainous fur coat.

Silver Dart was pulled to face into the light breeze, the prop was cranked and McCurdy decided it sounded good. Signalling the crowd to clear the way, he gunned the new 35-h.p. Curtiss motor, humped and lurched over the ice, and finally up into the air.

A crowd of boys and men on skates who were mostly doubtful he could fly followed him. "It was amusing to look back and watch the skaters—they seemed to be going in every direction—bumping into each other in their excitement at seeing a man actually fly," McCurdy said in an interview with the CBC in 1944.

Green's biography described *Silver Dart* lifting to the "frightening height of thirty feet" and flying for three quarters of a mile at about forty miles per hour. McCurdy, who had intended a test run only, turned the craft around (a great accomplishment at that time) and came back to his original starting place. He made a perfect landing as the crowd cheered and threw hats and mitts into the air. (The *Washington Star* mused that fifty percent of the watching crowd had names beginning with "Mc", and many were "McDonalds", an observation that can be made to this day).

The crazy contraption flew! Whooping and hollering. What a thing! The people of Baddeck lived in isolation, had to steam over the water at eight knots to get anywhere. The act of flying itself, of soaring away from the ground and being free, in a several-miles-an-hour age, was astounding, New Age, almost poetic. Poets enjoy seeing the meaning of things, and I like to think that Earle Birney, one of Canada's great ones, was anticipating the year 2009 when he wrote in *North Star West*:

Yet for a space we held in our
Morning's hand
the welling and wildness of Canada,
the fling of a nation
We who have ridden the wings of our
people's cunning
and lived in a star at peace among
stars
return to our ferment of earth with a
memory of sky.

In the next few days, several more flights were made with no problems, one for 20 miles over the area, another for 12 miles. That winter, McCurdy flew above a measured nine-mile course at Baddeck for a total of 1,000 miles.

The flight of the *Silver Dart* in 1909, conducted in that trial-and-guess period of aviation, was a triumph of relative ignorance. Canadians were excited when McCurdy flew the powered contraption off the ice. Since their history-making, but non-public, first powered flight in 1903 at Kittyhawk, North Carolina, nothing much had been accomplished by the Wright brothers, Wilbur and Orville.

Other aviation entrepreneurs were outpacing them. The Wright Brothers' aircraft had no wheels, and its low-powered engine needed the help of a catapult to get the aircraft airborne. Flight control was uncertain.

The press had almost ignored the Wrights, their feat wasn't mentioned in a scientific publication for several months. The prestigious *Scientific American* didn't get around to mentioning the Kittyhawk flight for three years. In fact, the *Scientific American*'s silver trophy for aviation competition first was won by an AEA member, Glen Curtiss. Posterity's verdict on the Wright brothers is that they were too secretive in protecting their invention, and rather poor businessmen.

So McCurdy's historic departure from the pull of Earth—he was ninth man to fly a heavier-than-air machine—was noted around the world. It was widely acclaimed in newspapers as far apart as the *Moose Jaw Times* and the *Washington Star*.

Perhaps some of the near-rapture at Baddeck was due to the spirit of change in 1909 Canada's population of seven million. The world was becoming smaller, urbanized, industrialized, consumer-driven, gender

Douglas McCurdy lifts into the air during his historic 1909 first flight of Silver Dart, skaters following on the ice of Cape Breton's Bras d'Or Lake.
Photo: Courtesy Airforce Magazine

Alys Bryant (left), just before her flight at Vancouver, B.C., on July 31, 1913. Amy Johnson (right).

WOMEN WERE DOING IT, TOO

THE GENDER-BENDER fact is firmly established: physique has little bearing on the business of flying, muscle or body size being second to quick thinking, keen reflexes, analytical application. Women fit the cramped fighter cockpit or the front office of the latest jetliner behemoth just as competently as men, and do so regularly in Canada. It didn't take long for Amelia Earhart to get airborne from Newfoundland and duplicate Charles Lindbergh's epic solo flight across the Atlantic. Alys McKey, first flight by a woman in Canada, 1913; Eileen Vollick, first woman to get a Canadian pilot's licence, 1928; Eileen Magill, first Canadian woman to fly across the international border, 1929; Nellie Carson, first woman to set an altitude record, 15,000 feet, 1930; Beryl Markham (flew the Atlantic east to west); Anne Lindbergh (many pioneering flights with her husband); Amy Johnson (record flight from U.K. to Australia); Jacqueline Cochrane (women's world speed record in a Canadian-built-and-powered Sabre jet); the hundreds of Canadian women ferry pilots and instructors during the Second World War; a thousand women in the same war who flew Yak fighters (Shturmoviks) and every other Soviet plane, including Lilya Litvyak, The White Rose of Stalingrad, with 12 victories and integrated, with many others, into a regular air force unit.

-conscious. Using cable technology, King Edward pushed a button in Colchester, England, and in a new tuberculosis centre in Montreal, the doors opened and the entrance was flooded with light, "much to the amazement of a crowd of dignitaries." The West was opening up and beckoning to areas of stagnant development; an advertisement in the East urgently pleaded that, "Thousands of nice girls are wanted in the Canadian West…Over 20,000 men are sighing for what they cannot get, Wives…If you cannot come, send your sisters." And, at Winter Harbor, Melville Island, an Arctic navigator unveiled a tablet on Dominion Day claiming the entire Arctic archipelago for Canada.

The Aerial Experiment Association in 1908 at Baddeck, N.S., left to right: Glenn H. Curtiss, Frederick "Casey" Baldwin, Alexander Graham Bell, Lt. Thomas E. Selfridge, J.A. Douglas McCurdy. *Photo: Parks Canada/Alexander Graham Bell National Historic Site of Canada*

THE IMPORTANCE OF BELL

J.A. DOUGLAS MCCURDY'S historic flight was the result of a unique organization controlled by Scottish-born Alexander Graham Bell, who had been interested in the concept of flight, and written about it, since the 1890s and would have a big influence in the early development of Canadian and American aviation. He and his wife, Mabel, had established a manorial home, *Beinn Bhreagh* (beautiful mountain), and laboratory buildings in 1891 on 1,600 acres with a commanding view of Bras D'Or Lake. Bell was conducting many experiments there with the determined aim of forging ahead of the Wright brothers, and employed

McCurdy and Frederick Walker "Casey" Baldwin, two recent mechanical-engineering graduates of the University of Toronto.

Earlier, McCurdy's father Arthur, editor of the *Cape Breton Island Reporter,* had left his newspaper to become Bell's secretary. Having lost two sons in infancy, Bell had a strong paternal feeling for five-year-old Douglas, wanted to adopt the motherless boy. His true father would never be far away. In 1893, they agreed on a trial adoption. But after one winter spent by Douglas at Bell's Washington home, his Aunt Georgina wanted him back with her in Baddeck. A Presbyterian, she wanted the boy to know the "dignity and discipline of work." So he returned home. Young Douglas liked to help out in the Bell laboratory, and rub shoulders with Bell associates. Bell later paid for McCurdy's education at the University of Toronto, where he graduated at age 20.

Mabel Bell could be called the First Lady of Canadian aviation. She suggested the idea of the Aerial Experiment Association to her husband and the two Canadians, according to the McCurdy biography. "Alex, you have some pretty smart young engineers here. And they're just as interested in flight as you are. Why don't we form an organization?" Dr. Bell, "was inclined to be too much of a dreamer…he was not even a good mechanic…Nor was he a good businessman, and it was his wife who compensated for that, she managed the immense estate." And she insisted on financing the project with her own money, a total of $45,000 (may- be half a million, today), to pay for equipment, and salaries.

The Aerial Experiment Association Ltd. was a typically Canadian project by wanting to be both Canadian and American, and was incorporated in 1907, at Baddeck, as well as Hammondsport, New York, in the Finger Lakes district south of Lake Ontario, for the duration of one year, with extension if decided. While the Canadian government rejected the practicability of flying and ignored experiments, the U.S. war department was aware of possibilities. U.S. defence planners already had rejected the usefulness of the Wright aircraft. Bell went directly to President Theodore Roosevelt, who released Lt. Thomas Selfridge, 25, the U.S. Army's leading aeronautical expert at the time. Bell also brought in Glenn Curtiss, 29, who was then focused on motorcycles and developing expertise at manufacturing light, compact and simple engines in Hammondsport. Uninterested in aeronautics although turning out engines for airships, he caught the spirit of Bell's team and embraced the idea of building airplanes. When he left the AEA, he would set up the first aircraft manufacturing company in the U.S. at Hammondsport that would become a legend of the industry.

So, the Aerial Experiment Association which would kick-start North American aviation spectacularly—a kind of private Free Trade Agreement—now comprised three Americans and two Canadians: Bell, 60, inventor (in Canada, he insisted) of the telephone, creator of the Bell telephone company, and one of the United States' wealthiest men; Lt. Selfridge, 25, Curtiss, 29, McCurdy, 21, and Baldwin, 25.

The five gathered at Baddeck, in 1907, and the details were worked out with the common aim, "To get into the air" in a practical airplane, with its own engine and carrying a man. Like musketeers, they swore "all for one and one for all." The minutes of their meeting show that it was agreed each member would design a machine, and that each would help the others in advancing their own ideas. Any inventions that resulted would belong to all of them. The projects undertaken were the *Red Wing* (Selfridge), the *White Wing* (Baldwin), the *June Bug* (Curtiss) and the *Silver Dart* (McCurdy). The headquarters of the association would be at Beinn Bhreagh. The agreement was signed, sealed and witnessed on Sept. 30, 1907, at the Halifax Hotel. In 1908, Glenn Curtiss made the world's first official flight in the Aerial Experiment Association plane *June Bug*, at Hammondsport, the immediate predecessor of the *Silver Dart*. McCurdy was the only other pilot to fly the *June Bug,* for 20 of its 54 flights and, in fact, it was the aircraft in which he taught himself to fly, according to CAHS, journal of the Canadian Aviation Historical Society.

They had begun by working on elaborate kites. Bell had earlier constructed a huge tetrahedral kite, *The Cygnet,* and it was hoped they would be able to launch it carrying both a man and a motor. A cumbersome method employing a barge-like craft carrying the kite, towed by a small steamer, succeeded in getting Lt. Selfridge to an altitude of 168 feet. But it ended up in the freezing water, a total wreck, the shivering, dripping Selfridge unhurt.

Back to the drawing board, this time at Hammondsport, and forward with Red Wing, a two-winger with a fixed stabilizer, a rudder at the rear, her nose boasting an elevator, and all standing on two runners. Her bamboo wings were covered with red silk. Curtiss had designed an air-cooled engine, V-eight and 40 horsepower. There was no schedule as to who would fly the experimental planes, and Casey Baldwin was the pilot at hand who gunned the engine, slid 150 feet over the ice of Lake Keuka, and lifted off to a majestic altitude of five feet and held it there for 319 feet before the engine quit. But Red Wing's first flight had been in nil-wind conditions. Five days later, there was a slight breeze that lifted one of the wings, and there was no control device to counter this. Casey Baldwin, again the pilot, emerged from the wreck unhurt. The fragile plane had been destroyed. The incident was given excited play in the newspapers.

Good can come of bad and, shortly after, the associates had solved the control problem by developing the aileron, an idea of Bell's. This was contested by the Wrights, who believed they had a monopoly on flight controls, and saw the Bell invention as infringing their unsophisticated and less practical wing-warping technique. A U.S. court decision was made in the Wrights' favour, and Curtiss had planned an appeal to the U.S. Supreme Court. But he dropped that idea when the First World War broke out and a "cross-licensing" deal was arranged. Ailerons were installed in all future Curtiss planes and became known in the U.S. as "the Curtiss control."

Several more planes were produced and finally the first water-cooled engine was made and fitted in 1908 into a design by Douglas McCurdy, the *Silver Dart* — so named because the wings were covered with a silver-coloured rubbery mixture. It was the most refined of the four AEA aircraft. Made of steel tubing, bamboo, wire and wood, it was flown several times at Hammondsport— the first public flight by an aircraft anywhere—and then was disassembled and moved to Baddeck the following winter. As noted earlier, the momentous first Canadian flight then took place on February 23.

In March, Bell addressed government elite at the Canadian Club of Ottawa. He set out to sell the idea that the Canadian government should act to recognize the new science of aviation, referring to airplanes as "aerodromes," as had scientist and aviation enthusiast Samuel Langley, before him. He had been an American for 35 years, he said, but McCurdy and Baldwin were Canadians, "and they want to go into the practical manufacture of these machines. They say, 'Cannot we do anything for the Canadian government?'" Editorials supported the two Canadians. But general opinion was divided on

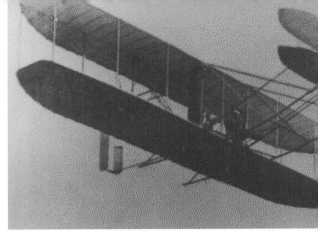

Early flight of Wright Flyer. The Wright brothers contested the Bell association's claim of having invented the aileron

the merits of airplanes. Rather than promising a new future, some said planes were a folly.

The Aerial Experiment Association was disbanded at the end of March, 1909, and McCurdy and Baldwin held the Canadian patent rights. The American rights were assigned to Glen Curtiss. Bell, who would end his aviation experiments in 1912, suggested that the young Canadians form an aircraft company and try to interest the Canadian government—and possibly other governments—in aviation. They could operate in one of his laboratory buildings at Baddeck, and so formed the country's first aviation firm, Canadian Aerodrome Company, which Bell financed. The machines were simply-made by today's standards, much the same way boys used to make balsa-wood-and-paper model airplanes, but with hardwood stringers and struts, covered with fabric and doped to make the covering tight and waterproof, wires attached and tightened to hold elements in place.

 Fine-tuning Baddeck No.1, first plane built in Canada, designed and flown by Douglas McCurdy at Petawawa army trials, 1909.
Photo: Courtesy Airforce Magazine

The brilliant inventor of the telephone had jump-started the aviation business in North America, with McCurdy in Canada, and Curtiss in the U.S., starting the first aircraft-manufacturing firms in their respective countries. Bell's contribution to aviation far outdistanced that of the Wright brothers.

MILITARY APPLICATION

WHO OF THOSE ON THAT lake ice watching the first Canadian flight would have known that almost at once minds would turn to advocating these novel, harmless machines as future instruments of war. The *Halifax Herald* reported in March, 1910, that a telegram had been received by Bell from General Allen, chief signal officer of the United States Army, expressing his "interest in the trials now being made at Baddeck."

Following approaches by McCurdy, Ottawa asked for a demonstration at Camp Petawawa, the 22,000-acre army base set up five years earlier along the banks of the Ottawa River, near Pembroke. On August 2, the two young men, bursting with eagerness to be entrepreneurs, brought out the *Silver Dart* before assembled officers. She made five flights, mostly at an altitude of 50 feet and at 50 miles per hour, demonstrating turns and stability. Everything went well until the last landing. The sandy soil was not as firm as the clay or sod the pilot was accustomed to at Baddeck, the motorcycle wheels sank too deeply, the aircraft went up on its nose and was demolished, ending the first military demonstration in Canada.

The army officers deemed the demonstration a failure and decided flying had no place in modern defence forces. Years later, in a belated attempt to associate itself more firmly with aviation, military authorities erected a monument where the Silver Dart crashed, proclaiming it the birthplace of Canadian military aviation.

But the demonstration had other significance: McCurdy and Baldwin had also produced the first powered aircraft designed and built in Canada, Baddeck No.1, and taken it as well to Petawawa. It was almost a copy of the Silver Dart, its tapered wings with dihedral on the lower plane, anhedral on the top plane, and larger wingtip ailerons. There was a slim fuel tank to fit within the wing, "an innovation which has since become standard practice and pioneered in the Baddeck No.1 (so named to show its Canadian derivation, as opposed to the Silver Dart, built in New York State)", according to Canadian Aircraft Since 1909. Baddeck No.1's demonstration at Petawawa ten days later was the first flight of a Canadian-made aircraft. The wingtip ailerons—now the oldest in the world—from Baddeck No.1 can be seen at the Bell Museum in Baddeck.

The *Silver Dart's* specially-designed engine was shipped back to Baddeck and eventually installed in a motor launch, which later sank in shallow water. It lay beneath the surface for more than two years before being salvaged. When its historical importance was realized, it was reconditioned and sent to the Aeronautical Museum of the National Research Council.

IN THE VANGUARD

HOW DID MCCURDY and Baldwin stack up in the list of those first to pilot an aircraft? The McCurdy biography shows that the Bell association's accomplishments were in the world vanguard: December 13, 1903, Orville Wright, American, Kittyhawk, N.C.; October 23, 1906, Alberto Santos-Dumont, Brazilian, Bagatelle, France; April 5, 1907, Louis Bleriot, French, in France; January 13, 1908, Henri Farman, English-born, at Issy-les-Moulineaux, France; March 12, 1908, F.W. (Casey) Baldwin, Canadian, at Hammondsport, N.Y.; March 28, 1908, Leon Delagrange, French, in France; May 19, 1908, Lt. Thomas Selfridge, American, at Hammondsport, N.Y.; May 21, 1908, Glenn Curtiss, American, at Hammondsport, N.Y.; May 23, 1908, J.A.D. McCurdy, Canadian, at Hammondsport, N.Y.; June 8, 1908, Alliott Verdon Roe (a huge influence on Canadian aviation, yet to come), English, at Brooklands, England.

Fifth and ninth men on the world list, Baldwin and McCurdy were the seeds of a Canadian aviation program that at first were denied a plot in which to grow. In contrast—partly because of the work done at Baddeck—Glenn Curtiss flourished in the U.S., founding the first aircraft manufacturing plant, and eventually would be tagged by some as "the Henry Ford of Aviation." In 1910, the U.S. Navy allowed him to fit a wooden platform over the deck of a cruiser to rig the first aircraft carrier. He established aviation schools, and his factory supplied land and water aircraft for the U.S., British and Russian forces during the First World War. Lt. Selfridge was killed in a 1908 crash of an airplane piloted by Orville Wright, yet still was honoured by his government when later an air force base was named after him.

McCurdy's first Canadian flight was one of a galaxy of aeronautical accomplishments. In the span of just two years, he and his colleagues would invent the aileron, a flight-control mechanism that was hugely significant and still vital to today's aircraft (until then, in trying to control their aircraft, the pilots sometimes pulled wires that altered the shape of the wing—"warping," initiated by the Wright Brothers—and achieved some effect, if lucky, that ailerons would always easily and safely provide; earlier aileron inventors are suggested, but none offered the control of the AEA's); the tricycle landing gear; pontoons—almost a definition of "bush flying," that made Canada's wilderness of lakes into thousands of landing places. He was the first to pilot a plane powered by a water-cooled engine; made the first long-distance ocean flight, Florida to Cuba; made the first flight on skis; was the first to demonstrate the idea of bombing from the air; flew the first figure-eight; was the first to communicate from an aircraft by wireless—in effect, the first celestial navigation signal (a 30-word message in Morse code for which he was jolted with an electric shock each time he depressed the key). And later, the determined and talented McCurdy became known as the "John the Baptist" force behind formation of the Royal Canadian Air Force. He had an important government aviation job during the war, and ended up lieutenant-governor of Nova Scotia.

Gilbert Grosvenor, then chairman of the National Geographic Society, wrote in 1959 that he had known Peary, Shackleton, Amundsen, Lindbergh, Byrd and, "I regard John A.D. McCurdy as a man who ranks with the very greatest of these." In 1910, McCurdy became the first Canadian issued with a pilot's licence, No.18 with the Aero Club of America. In 1959, the fiftieth anniversary of his Baddeck flight, he was awarded the McKee Trophy

In the world's first "public" flight of an aircraft, Douglas McCurdy coaxes the *Silver Dart* higher above Bras d'Or Lake in Cape Breton, at Baddeck, N.S., on Feb. 23, 1909. *CF photo*

for his contributions to the advancement of flying and Canada.

Anyone who's been around the block probably understands that observing Canada's centennial of powered flight wasn't only to bask in the past, and maybe ask questions, but also to see possibilities. Canada's 2009 Centennial of Flight was an observance of achievement. Or, if your point of view is perverse, that first flight also marked the first day of a disdainful, hands-off policy of a long line of leaders in Ottawa who left stillborn the ambitions and progress of both civil and military aviation many times through the coming one hundred years. •

 From the cockpit of his favourite Camel fighter plane, a deadly glance from William Barker, "VC, D. S. O., M. C., etc.", most-decorated Canadian ace of aces in Britain's Royal Flying Corps.

CHAPTER TWO

THE FIRST WORLD WAR

ON A NIPPY FEBRUARY DAY at CFB Trenton, in 1956, we had finished the landlocked School of Instructional Techniques, and I was stick-jockeying my way through the syllabus of Flying Instructors' School. Our T-33 trainer had lifted off the long and extremely busy runway when the instructors' instructor drawled that we would fly over to a little runway at Deseronto to practise emergency landings and such. During the First World War, smaller airfields were built in southern Ontario, such as Rathbun (the former farm owner's name) and Camp Mohawk (Deseronto), on an Indian reserve, near Belleville; and at Long Branch, Leaside and Armour Heights, in the Toronto rea. All were satellites of Camp Borden.

Well! Mention of Deseronto brought to mind Biggles—hero of young readers' adventure books by Capt. W.E. Johns—and other mental litter to pump adrenalin into the imagination, calling up visions of those snarky Royal Flying Corps tunics, and two-winged "Flying Jennys" zooming and twisting, landing shakily on the Deseronto runway. Maybe this preoccupation influenced the antics to come.

After three years flying Sabres in Europe, I felt pretty capable as we began to practise touch-and-goes. Third time around, the instructor turned off the power boost for the control surfaces so I could experience man-handling the aircraft to a landing. That accomplished, he told me to overshoot. I realized that the boost hadn't been restored. But I figured he must know what he's doing. While climbing away, he yanked back the throttle, drawled "flame-out," and told me to execute a dumb-bell—climb steeply, bank sharply, reverse bank and bring the aircraft around 180 degrees for a dead-stick landing back in the opposite direction.

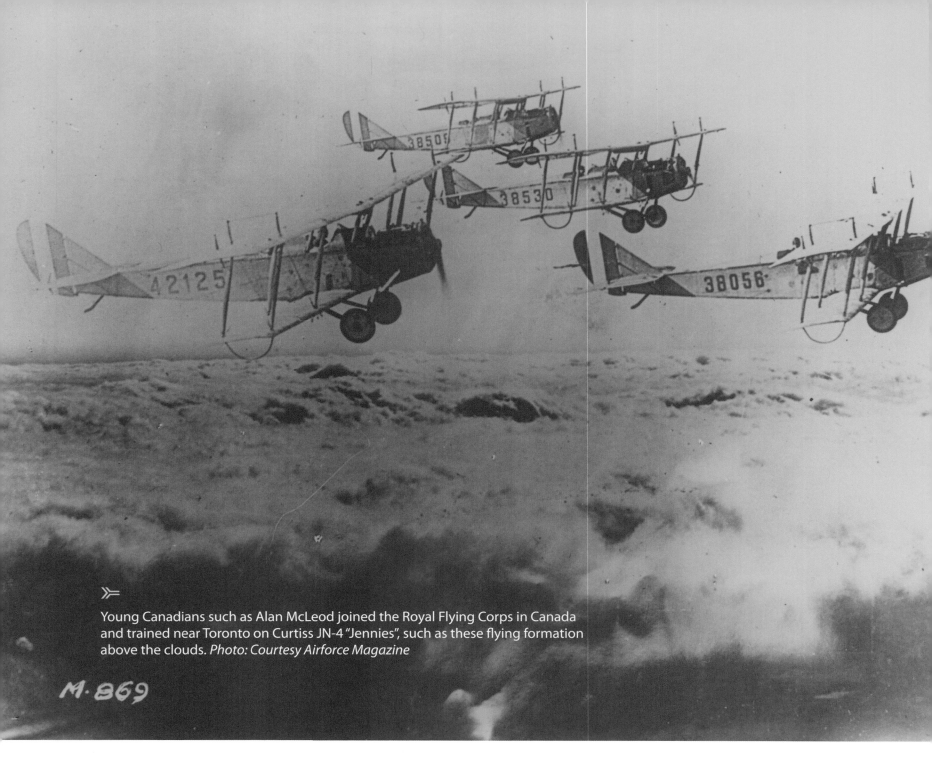

Young Canadians such as Alan McLeod joined the Royal Flying Corps in Canada and trained near Toronto on Curtiss JN-4 "Jennies", such as these flying formation above the clouds. *Photo: Courtesy Airforce Magazine*

Okay! But hard work. I was at the apogee and had struggled half way around sure I would make it when I heard him grumble, and felt him grab the stick. A horrified expletive as he realized our muscles still were working against the slipstream, without hydraulic help. He turned the boost back on, completed the manoeuvre, said nary a word more and flew the ship back to Trenton. I could tell that if he had his way, I'd be bounced off course right then, no posting to advanced flying school at RCAF Station Gimli to instruct NATO students from Canada, Britain, Turkey, West Germany, Netherlands, Greece.

While I mused my fate, I was pretty sure that whatever he did to me, it couldn't be as bad as what had awaited some of those poor sods in RFC uniforms when they got over the Western Front.

Line-up of Sopwith Camels ready for the Western Front. This was the favourite fighter aircraft of William Barker, VC, a Canadian who flew with the Royal Flying Corps. Dangerous to fly, but loved by those pilots who mastered them. *Photo: Courtesy Airforce Magazine*

IN THE AIR OVER VIMY RIDGE

IT WASN'T ALL DOGFIGHTING in the clouds for Canadian airmen in British uniforms. The First World War battle for Vimy Ridge in France was seen as a defining military action in Canada's approach to true nationhood. Fought during "bloody April" of 1917, the Canadian victory ended a trench-war stalemate which had killed hundreds of thousands of soldiers, and pointed the way to war's end. Canadian pilots with the Royal Flying Corps were in the thick of the battle for Vimy Ridge.

In fact, RFC planes flew back and forth in the midst, and regardless, of the artillery barrage they were master-minding as the army's "eyes," the sky thick with shells that pilots sometimes saw coming and going. Some of their aircraft were hit by their own bombardment, as well as enemy fire. Being a corps of the British army, the RFC had a multitude of missions to carry out before and after the April 9 offensive began: reconnaissance, artillery co-operation, contact patrols, bombing of enemy communications, strong-points, and troop concentrations.

German scout planes of the Fokker and Albatross class had out-performed many British aircraft, such as the Martinsyde "Elephant," the FE-2d and the DH-2, and a new generation of planes was thrown into the Vimy fray, including Bristol Fighters, SE-5s and Camels. Technology was constantly changing aerial combat: like the fashion parade, today's favourite plane was tomorrow's cast-off. This air war was shaped as airplanes and armament developed, and no other air war would be like it again. As with the whole idea of flight, the conflict was pioneering, vivid with bravery and made notable by the skill and daring of intrepid airmen.

During the British artillery bombardment, "There seemed to be millions of spots of flame spread all over the countryside, and it must have been as near hell as possible for those down below," a pilot was quoted in *The Royal Flying Corps in World War 1*. It was equally hell for those up above, he said. The air was stiff with flying shells, their shock waves bumping the aircraft as they passed under and over.

FLYING OVER VIMY

THE VIMY MONUMENT is considered by some to be one of the most magnificent memorials ever erected. The site on which the Canadian Memorial at Vimy Ridge sits, as well as the surrounding 100 hectares of land, was given in 1922 out of gratitude for sacrifices made by Canada in the "Great War." The people of Canada built the memorial as a tribute to the countrymen who fought there, particularly to the more than 66,000 men who lost their lives. From the monument, other places where Canadians fought and died are visible. More than 7,000 are buried in 30 war cemeteries within a 16-kilometre radius. On the outside face of the enclosing walls are the names of 11,285 Canadians killed in France whose final resting places are unknown.

Walter Seymour Allward designed the monument. He was an unusual Canadian artist in that he was both an architect and a master sculptor. His design was selected in a competition in the early 1920s. While impatient authorities fretted, Allward spent more than two years roaming the quarries of Europe, determined to find the stone that met his exacting standards. Work began in 1925 and, 11 years later, was unveiled by King Edward VIII. The base and twin pylons contain almost 6,000 tonnes of an extremely durable limestone from Yugoslavia (present day Croatia). The 20 sculptured figures that grace the monument were carved where they stand from huge blocks, including those at the top of the twin towers. The carvers used half-size plaster models sculpted by Allward, and an instrument called a pantograph to reproduce the huge figures to the proper scale. A master carver added the finishing touches.

The twin pylons tower 27 metres above the base of the monument. Because of the height of the Ridge, the topmost figure—that of Peace—is approximately 110 metres above the Douai Plain to the east. Allward said that the pylons represent Canada and France—two nations beset by war and united to fight for a common goal—peace and freedom.

Flying my F-86 Sabre fighter out of Marville almost 40 years later, I chanced upon the towering Vimy Ridge memorial. From my cockpit, the sight of the pylons poking the skies from the former field of battle kindled memories of what I had heard and read through the years. Emotion built and I pushed the stick forward, airspeed building as I dove. Levelling out, the mach-metre registered near point nine, about 600 knots, and I let the Sabre settle to just above ground level. As the monument grew huge in front of me, I pulled back, climbed vertically and rolled the aircraft in a salute, to about 12,000 feet (3,900 metres).

On the way up words rolled past in my mind, the few words I could remember from the Act of Remembrance: "They shall grow not old, as we that are left grow old…We will remember them…."

18-year-old Alan McLeod after winning his flying badge (wings), the hero's medal, the Victoria Cross still to come. *Photo: Courtesy Airforce Magazine*

FROM BADDECK TO BISHOP

CANADA'S FIRST WORLD WAR air record is replete with tales, photos, movies, paintings, old medals, monuments, gravestones, maybe a few lies, feats of astounding bravery and heroics, dogfights over the Western Front, flamers, no parachutes, bi-planes and tri-planes. Some of the names: Billy Bishop, William Barker, Roy Brown, Wop May, Punch Dickins, Fred Stevenson and, yes, Baron von Richthofen, still subject of a debate about whether or not a Canadian, Roy Brown, ended the German flyer's illustrious air-fighting career.

Here's what the Canadian government did during the 1914-1918 air war: after shipping the Canadian Aviation Corps—comprising three men and a contraption posing as a warplane—to England, where all disappeared forever, the military and government at home denied that "aviation" was even a word. Wrote K. M. Molson in *Canadian Aircraft Since 1909*, "From then on during the war the Canadian Government stoutly resisted any involvement in this new form of military activity until the summer of 1918. This non-involvement even included refusing to participate in the training scheme set up in Canada by the RFC (Royal Flying Corps, renamed Royal Air Force in 1918) although the Government finally agreed to hold the mortgage on the Curtiss Aeroplane Co's. factory, which was to supply training aircraft for the RFC (Canada)."

Meanwhile, Douglas McCurdy, after the events at Baddeck, occupied himself with a number of activities aimed at promoting Curtiss's aircraft business, and juicing up interest in aviation. Along came the First World War, which McCurdy had predicted in speeches. In Toronto, McCurdy argued that there was no reason why there shouldn't be a squadron of at least twelve planes in Canada to assist the militia. He wanted the government to send some machines over to the Allies. He offered patriotic suggestions, but Militia Minister Sam Hughes rejected his overtures. As time went on, it became increasingly evident that whatever McCurdy would do to further the idea of an air arm for the Canadian military effort, he would do without government help.

THE CAC DEBACLE—HUGHES' FOLLY

NORMAN SHANNON DESCRIBED Minister of Militia Sam Hughes's goof with the three-man Canadian *Aviation Corps in Airforce* Magazine, in 2001. He wrote that Hughes was approached at Valcartier army base near Quebec City just before the Canadian expeditionary force sailed for England in 1914, with the suggestion he form an air force. It appeared to be Hughes's war to run, and he ran with Ernie Janney's idea—a man from Galt who had never flown, maybe never even seen an aircraft—installing him on the spot as a Captain and provisional head of the Canadian Aviation Corps, neglecting to inform his deputy minister or the defence department.

With authorization to spend $5,000, Janney rushed out to buy a uniform, a handgun to strap on and, at the Burgess plant in Marblehead, Mass., a biplane with a wingspan of 46 feet, and a fuselage mounted on a pontoon (a flying boat). This he arranged eventually to have flown from Lake Champlain into Canada. While this was going on, Hughes had added William Sharpe, who had flown, and Harry Farr, an engine man, to the aviation corps. They all sailed away to Europe.

Gen. Sam Hughes, disgraced Canadian militia Commander, thought the war was his to run. *Photo: Courtesy Airforce Magazine*

But few in Canada knew the Aviation Corps had left, wrote Shannon, "while none in England knew it was coming, or cared…the Burgess-Dunne remained behind on the dock at Plymouth until the dock-master insisted that it be moved or he would throw the "pile of junk" overboard. The Burgess-Dunne and the Aviation Corps became lost in war's twilight zone as Sam Hughes forgot all about it."

A year later, Hughes told young Canadians who wanted to fly as observers for the army that if he needed to observe the enemy on a battlefield, he would climb a tree. "Janney wandered around England for three months, ended up back in Canada, Sharpe became Canada's first aircrew casualty when he died in a flying accident while serving with the Royal Flying Corps, and Farr was discharged within three months," after the CAC was disbanded.

So he added "McCurdy Flying School" to the sign at his Curtiss Aeroplane factory. And the British came to his rescue by throwing business his way. Probably because it cost nothing and didn't require any effort on its part, the Canadian militia department gave McCurdy permission for a "private school" for recruiting and training in Canada of candidates for the Royal Naval Air Service. Canadian recruits would get preliminary training, then go to Britain for finishing. Candidates paid $400 for instruction, most of which was refunded by the RNAS, along with uniform allowance and pre-paid second-class passage overseas. Then the RFC joined the plan. The rivalry between the two British services was so intense, McCurdy had to set up separate schools, the navy men at Toronto Bay, using Curtiss Flying Boats, and the RFC at Long

Bungled Canadian Aviation Corps authorized in 1914 was three-men strong and accompanied an unwarlike Burgess-Dunne tailless aircraft to Britain, never to be used. The CAC was quickly disbanded. *Photo: Courtesy Airforce Magazine*

Branch Rifle Range's hastily-constructed flying field, on Curtiss biplanes, JN-4s, with in-line engines.

JIM SHILLIDAY | 33

BE 12 was used to observe enemy troop movements.

The top fighters overseas used rotary engines, so graduates had to take a transition course on Avro 504ks after arriving in England.

Far from being a daredevil's prank, it suddenly became a noble thing for one to frolic in the heavens, and Toronto papers of 1915 gave dozens of detailed accounts of the glamorous happenings at the flying school. *The Toronto Star* reported:

'You'll do,' said pilot to pupil after just three flights.

The phenomenal record of learning to manipulate a flying boat in 118 minutes' experience goes to Edward Blumfield Waller, of 192 Jarvis Street, a student of the Curtiss Aviator School. Young Waller began to fly only last Wednesday, and passed his final test in the hydroplanes yesterday.

On his third flight, which was made early last Thursday morning, Waller noticed the pilot fixing his goggles with both hands, and was surprised to learn on reaching the hangar that he himself had been in complete control of the machine during the entire flight.

As the Western Front ate up soldiers and airmen, Britain asked if it could formalize things by setting up RFC Canada to build and manage flying fields in this country, and to recruit and train young Canadians for their units, and for the Royal Naval Air Service. Soon, Canadian workmen in the middle of 1917's frigid winter were working at an army base called Camp Borden, just 70 miles north of Toronto, topping rectangular wooden structures with rafters they had pounded together and manhandled with poles into position to form distinctively-shaped, carapaced roofs. After six weeks, 15 "flight sheds" (hangars) were strung out in an imposing arc beside the landing strip. Smaller airfields were built around Toronto, and farther east at Deseronto, near Belleville, but Camp Borden, wrote historian S. F. Wise, in *Canadian Airmen in the First World War*, "was therefore the origin of what was later to become the senior air station of the RCAF."

Another aviation historian of the time, Normand Marion, wrote that Camp Borden was "one of the finest aviation centers in the world." The hangars each could hold 10 Curtiss JN-4 aircraft—made in the Curtiss plant in Toronto at a rate of ten a day, and purchased by the RFC—and had sliding doors at each end. Forty-two other buildings went up by summer, and flying training had begun. Airmen flew at the field that winter and "proved to the world that cold weather flying was possible, given a minimum of precautions and attitude adjustments."

A basic flying training system was adopted, a comprehensive syllabus which has changed little to this date, according to the summer 2004 issue of *Airforce* magazine. "It included dual instruction and solo practice on the effect of the controls with power on and off; straight and level flight; climbs and turns; misuse of controls in turns; sideslips; stalls; spins; slow flying; take-offs; approaches and landings; forced landings; and advanced manoeuvres. Formation flying, cross-country flights, and aerial practice of the photo and bombing skills that they had learned earlier completed the initial flying training phase."

At Camp Borden, not one student was killed or injured during the McCurdy schools' operation. RFC Canada training started in April, 1917, and thousands of flying hours were recorded on board the American-built "Jenny," The Curtiss JN-4, and on the Canadian version, dubbed the "Canuck". But with the mediocre standard of instruction at some of the other schools, the touchy engines and fragile aircraft, not all flying in the system went smoothly, and by the end of September, 32 planes had been destroyed at Camp Borden alone, seventy-one in all RFC Canada. Camp Borden had the first fatal flying accident in the history of military aviation in Canada, when Cadet J.H. Talbot was killed on April 8, 1917.

Sam Hughes and the Canadian cabinet still rejected the idea of a separate Canadian air force. After the flying-training schools had been in operation for six months, the government's frigidity toward aviation was being condemned in newspaper editorials. "It appears," said *The Toronto Star*, "that one cause of this discrimination against aviation students is the fact that the Canadian Department of Militia has hitherto failed to recognize aviation as a branch of the Canadian Army, although it has been demonstrated since the first day of the present conflict that for modern armies an aviation branch is absolutely essential. The aviators are now described as the eyes of the army." During the early part of the war, the RFC flew in support of the British Army, in the role of artillery co-operation and photographic reconnaissance. This work gradually led pilots into aerial battles with German pilots, and later in the war included the strafing of enemy infantry and emplacements.

Hughes had been dismissed from the Canadian army during the South African War, Ronald G. Haycock wrote in *Sam Hughes: The Public Career of a Controversial Canadian, 1885-1916*, for "military indiscipline and public exposure of incompetent British generalship." In 1911, the fast talker seemed to have redeemed himself and was given the militia portfolio in R.L. Borden's new Conservative government. During the First World War, he rejected the idea of French-speaking army units and, in 1916, was fired for "favouritism, confused civil-military functions, disrespect of Cabinet, administrative incompetence and scandals such as the Ross Rifle (frequent jamming)…his erratic talents never matched the demands of high office during total war."

TRAINING FOR WAR

IN THE EARLY DAYS—at some Canadian schools and on RFC squadrons in particular—flying training was somewhat similar to the early life of an emperor penguin: time in the egg, a few months sitting on Daddy's feet to keep warm, then, waddle to a cliff of ice and dive in to become a "flyer."

RAF Air Marshal Sir Hugh Walmsley later succinctly recalled his experience with British flying training: "Our so-called instructors were not flying instructors at all." What did he mean? A future RAF chief of the air staff, J.C. "Jack" Slessor explained that after paying for instruction at private flying schools in England and getting a Royal Aero Club certificate, RFC candidates reported to Central Flying School for instruction to earn their badge, or "wings." This was perfunctory, Slessor said, "…dual instruction (pre-war) consisted of a few circuits and landings, in which the instructor flew the aeroplane, with the pupil having his hands and feet on the controls 'to get the feel of them'. An occasional bang on the back to attract the pupil's attention revealed the instructor waving his hands to show that he had relinquished control. Afterwards the pupil was asked whether he felt confident in himself, and if he answered 'Yes' he was sent off solo." The pre-war service atmosphere, Slessor recalled, with pilots allotted their own machine and, "flying wherever they chose, landing where they could, soon bred skill and confidence."

Lack of a comprehensive syllabus increased the danger to student pilots. There was no standard manual for instructors, no aptitude tests, and no instructors' schools. Most of the men recruited as instructors were home from the front "on rest." Charles Portal, one of the RAF's most distinguished chiefs of staff, was "astonished at the ignorance of the majority of his instructors." Vernon Brown, who became a leading test pilot and accident investigator, learned to fly in a Vickers Gunbus. "The machine had no dual control, so the instructor sat on the fuel tank behind Brown with his hand on Brown's shoulder. 'Right rudder, you fool!' Brown would hear him shouting...but Brown passed his test. 'You appear to be a perfectly good pilot,' he was told at CFS. 'You can be an assistant instructor here.' With fifteen hours solo on various types he started teaching others to fly."

The need for pilots became so urgent that below-average pupils might be given the benefit of the doubt, often with tragic result. Hours of instruction were cut, trainees sent solo before they were ready. There were so many accidents pupils had to pay for the purchase of wreaths for those killed...Trainee pilots awaiting advanced instruction in combat tactics, air firing and formation flying were liable to be diverted overseas to fill empty cockpits, the intention being that the squadron should train them...Many flew into battle with no more than five hours on type.

Considered by many to be Canada's greatest air ace with the RFC, William Barker of Dauphin, Manitoba, was an observer before becoming a pilot. Observers seemed to pick up flying technique by association and, since there were dual controls, often landed their aircraft (as Barker did) when the pilot was out of commission. Barker mastered the dual-control S11 Farman "Shorthorn" and was reported to have soloed after just two flights—one of 50 minutes and one of five minutes. Wrote his biographer, Wayne Ralph, in the deeply-researched *Barker VC*: "In the days before lesson plans and structured syllabus, a typical dual instruction flight rarely exceeded 50 minutes, and many flights were 20 minutes or less. Learning was more by osmosis or intuition, than through explanation, demonstration, imitation, and correction...Since few (instructors) had any background in teaching, the calibre of instruction varied from reasonable to awful...Talented men like Barker suffered less from poor instruction...a belief held by many instructors was that a couple of accidents were a positive learning experience...."

Douglas McCurdy was happily turning out pilots when, abruptly, the British Army ruled that it would stop using Canadian flying schools. Canadians who wished to fly with the RFC would take all of their training in Britain. The Royal Navy decided the same. This cost McCurdy. Three hundred students had not yet finished training. He refunded their fees and closed the school. He was proud of what he had accomplished, graduating 300 of what he considered the finest airmen anywhere. Time would endorse his pride: among his grads were Raymond Collishaw, W.A. Curtis, Robert Leckie, all future Canadian air marshals, and a Hammondsport graduate was Henry H. (Hap) Arnold, future general of the U.S. Army Air Force during the Second World War.

McCurdy went to see Prime Minister Borden in the spring of 1916, doggedly promoting an official Canadian air-training program. Borden gave him a signed letter addressed to the Canadian High Commissioner in London. During his trip to England, he talked the British into compensating him by letting the RFC take over his training school. English staff picked up training chores and opened other schools. McCurdy then was given an order to build a two-engine, 80-mile-an-hour bomber for the Royal Navy. So he built the world's first twin-engine bomber. By war's end, he had built 13 of them, and 1,210 JN-4s (Canadian), which had become the official training plane of the empire, and the first Canadian-built airplanes exported to the U.S, making up half the training fleet in that country. No other combat aircraft were built in Canada until 1938.

Since the United States paid more attention to the possibilities offered by aviation, McCurdy engaged in several business ventures there. He made experimental flights in planes and flying boats for Glen Curtiss in the Long Island area in 1912 and 1913, and busied himself learning the manufacturing methods at the Curtiss factory in Hammondsport, New York, which was producing the majority of aircraft and engines made in the U.S.

(With the war over, McCurdy would turn to manufacturing civilian aircraft. By the mid-twenties, trans-Atlantic flights were common enough to convince him that commercial air travel was the coming thing, and in 1931 he took over the bankrupt Curtiss-Reid Aircraft Ltd. and became its president).

Konnie Johannesson with an RCAF de Havilland DH60 Moth in 1936, when he was a Flying Officer in the RCAF Reserve.

RFC INSTRUCTOR IN EGYPT

KONRAD (KONNIE) JOHANNESSON, of Winnipeg, enlisted in the 223rd Overseas Battalion of the 2nd Brigade of the Canadian Army in March, 1916, and in April, 1917, went overseas to England, eventually ending up in the Royal Flying Corps as an instructor at the RFC airfield at El Khanka, Egypt. In a biographical sketch, he wrote: "Up to now, flying training followed no set pattern, each Instructor used his own method and consequently there was a lot of wasted time in the air. Some would spend hours on landings, with the pupil only in control on the actual landing. Hence the pupils could make beautiful landings consistently but were hopeless in the air, and would not even venture a turn to the right. Anyhow, it was not the kind of training that 'active service' pilots should have. From this hopeless inadequacy in training developed the need for a standardized form of instruction. And so, the 'Gosport system' was inaugurated in the formation of the Flying Instructors School (F.I.S). This school was located on a new 'drome just a few miles South of Heliopolis."

War is never fair, and the chance of being put "in harm's way" is just that, luck of the draw. Some pilots, who had thankfully escaped the muddy trenches and then combat, such as Johannesson, could wonder about the levelness of the playing field: "All the while the war was raging in France and the Middle East, news would reach us of former friends and

A Laird 740, CF-APY, taken in front of Johannesson Flying Service, Winnipeg Airport, in 1939 or 1940. *Photo: Courtesy Brian Johannesson*

acquaintances who had made the supreme sacrifice. We would drink a silent toast to their memory in the Mess, and recall our association and experiences with them. I am sure that each and every one of us present on these occasions felt guilty in some way, by reason of our "plush" assignment to School Flying. We would have no doubt had the opportunity to do our bit on Active Service, had the war lasted long enough."

As the war progressed, training became safer. But, by the end of the war, learning to fly had killed hundreds of pupils. The RFC increasingly drew on men from across the British Empire, including South Africa, Canada and Australia. More than 200 Americans joined the RFC and learned to fly in Canada before the U.S.A. became a combatant. Eventually Canadians made up nearly a third of RFC aircrew. Canadian pilots have always been at the top of the heap in air war. The proof was in the numbers. Britain's RFC would have been relatively ineffective without Canadian airmen: 171 of the 863 known British Empire aces of the First World War were Canadian; of the 25 British aerial fighters who shot down 30 or more of the enemy, 10 were Canadians.

THE THREE BS—BISHOP, BARKER AND BROWN

THREE CANADIAN FIGHTER pilots in British uniform—two of them Victoria Cross winners—came out of the First World War with a combined air combat "hat trick": highest number of aerial victories; most-decorated; and vanquisher of Germany's Baron Von Richtohfen, The Red Baron. Billy Bishop, from Owen Sound, Ontario, shot down the most enemy planes, 72; William Barker, of Dauphin, Manitoba, downed 50 planes, had the most medals (including six presented by the King of England at one time); and Arthur Roy Brown, from Carleton Place, Ontario, had 10 victories, including the most famous air battle of the war, the downing of Manfred von Richthofen.

William Avery "Billy" Bishop, first Royal Flying Corps winner of the Victoria Cross, enlisted in the 4th Battalion, Canadian Mounted Rifles in 1914, after attending Canada's Royal Military College. He transferred to the RFC and, as a fighter pilot in France at age 24, developed a reputation for being a "lone wolf," sometimes leaving squadron formations to stalk alone, often spending more than seven hours a day in the air. In 1918, he destroyed 45 enemy aircraft in five months, 25 of them in a 12-day period. A legendary marksman, Bishop took off in his Nieuport on dawn patrol during an army offensive, wearing pyjamas under his flying suit, and launched a lone attack on an enemy airfield near Cambrai, well behind the lines. He shot down four aircraft, returned to base with more than 100 bullet holes in his aircraft, many from ground fire. He was gazetted for a Victoria Cross, while still awaiting a Military Cross and a Distinguished Service Order. Although it was not usual for unverified claims to be recognized, he was backed up by superior officers. Combat reports, squadron records and official communiqués supported several other unverified claims. This exposed him to accusations in later years that he was a fraud and had inflated his victory claims. During the Second World War, Bishop was given the honorary rank of air marshal, and served as director of recruiting for the Royal Canadian Air Force. He died in 1956.

William George Barker was an infantry machine gunner in France in 1915 and, like Bishop, transferred to the RFC and became a fighter pilot. His combat mission on October 27, 1918, that earned him a Victoria Cross—the second last of the air war—became a *cause celebre* for

British and Canadian military and newspaper people looking for heroes. Flying and fighting for his life against large groups of hostile aircraft in five separate air battles—and despite terrible gunshot wounds—Barker, 24, managed to shoot down four of the enemy. The world was told he had engaged sixty enemy airplanes single-handed, a claim he never made. He was more proud, his biographer claimed, of his broad war record. Wrote Wayne Ralph in 1997: "…Canadians seem to like their heroes *smaller* than life…Almost all the magazine articles and narratives about Barker in the past 20 years have been written by British or American writers who still like their heroes *larger* than life, and don't mind them having a darker, more complicated side." Novelist Ernest Hemingway included a portrait of Barker in *The Snows of Kilimanjaro*. A daring, dedicated pilot with a deadly gun eye, Barker fought in the air for longer than most, amassing a total of 50 victories, 46 of them in the same scout plane, a feat never matched, anywhere. In 1924, Barker helped found the Royal Canadian Air Force and started one of Canada's first commercial air services with W.A. Bishop. He was killed in 1930 in an air crash at Ottawa's Rockcliffe Air Station.

Arthur Roy Brown, a 24-year-old naval squadron flight commander, was flying a Sopwith Camel when he scored his final victory of the war, his tenth, on April 21, 1918. His flight encountered Jasta 11, an enemy squadron looking for action. His combat report: "Dived on a large formation of fifteen to twenty Albatross scouts, D.5's, and Fokker triplanes, two of which got on my tail, and I came out. Went back again and dived on pure red triplane which was firing on Lieutenant Wilfrid May. I got a long burst into him, and he went down vertically and was observed to crash by Lieutenant Francis Mellersh and Lieutenant May. I fired on two more but did not get them." Brown was officially credited with

The Three Bs: Billy Bishop, left, William Barker, middle, and Roy Brown, right, all made combat history over the western front. The Victoria Cross, upper right.

shooting down the Red Baron, ending the German's victory score at 80 British aircraft, 123 pilots or crewmen captured or killed. On viewing Manfred von Richthofen's body, he said, "If he had been my dearest friend, I could not have felt greater sorrow." For this action, he received a bar to his Distinguished Service Cross. But during the air battle, a group of Australian Field Artillery machine-gunners had been firing at Richthofen's Fokker Dreidecker 1, as well, and claimed credit. The debate goes on still. Brown returned to Canada where he worked as an accountant, founded a small airline and became an aviation magazine editor. During the Second World War, he entered politics after his application to join the Royal Canadian Air Force was rejected. He ran for Parliament in1943, but lost, and died the following year.

William Barker examines one of his victims on the Italian Front.

TEENAGE TIGER WINS VC

WARTIME AVIATION CREATED many heroes and one of the greatest was Alan McLeod of Stonewall, Manitoba. McLeod learned to fly in one of the JN-4s made in Douglas McCurdy's Toronto factory. As did all young Canadian recruits for the RFC, he began his training at Burwash Hall, University of Toronto. There they were indoctrinated and given basic military training, followed by brief class work at the School of Military Drill and Discipline. In June, 1917, the student pilots were moved to Long Beach, west of Toronto, to start flying training. The country boy was airborne for the first time on June 4, and by June 9 had a total of six flights and 135 minutes of flying instruction, and was sent solo. Eight days later, with five hours and 10 minutes solo time logged, he was ordered on to Camp Borden to be taught military flying.

A BRAVE BAND OF BROTHERS

THE CANADIANS WHO FLEW with British squadrons were thought of as part of Canada's air war heritage. But our country had no such official background because the government had rejected the idea of air power. So, when the Second World War broke out, authorities had to dazzle Canadians with their footwork by issuing a booklet, Canada's Air Heritage, to be distributed to aircrew graduating classes in order to give them a feeling of having roots. Additionally, Billy Bishop, William Barker, and a third VC winner, Alan McLeod, were painted in oils and copies distributed to air force stations across the country. Many of these are still to be found in air force buildings, Canadian Legions and museums.

McLeod didn't like Camp Borden. It was a service training station, and that meant, "Do as we tell you," and he was notorious as a stem that didn't bend in the wind. A tall, strong boy, he had driven his father's Ford at a break-neck 30 m.p.h. along rural ruts, and frequently was called before Stonewall Collegiate's trustees for behaviour they questioned. He was known as a "caution" around town; "What will he do next?", but always with a chuckle.

Are heroes born, or are they a product of their environment? A local history book tells how young McLeod was returning from school when he saw a dog hobbling with a trap on its foot, and was unable to get near the animal to set it free. He followed it a long way into the countryside, finally succeeded in removing the trap. The Argus newspaper editor predicted, "a boy showing such fine qualities will be heard from later in life." In another local history, high school friend Anne Taylor:

"Walking down the main street of Stonewall in the early spring of 1917, I saw a big bay horse coming towards me at a sharp canter. The rider was Alan McLeod and he pulled up beside me and dared me to ride the horse. The look in his eye warned me to be wary. If Alan enjoyed riding that horse, then it must be a big-spirited beast. Alan enjoyed pranks and jokes but they were always good-natured. Although we were in high school, he was running one year behind because he had insisted on taking Grade Seven over again. He said he owed it to his father who believed

in things being done thoroughly, and Alan had a difficult time applying himself to school. Alan was known to most of his school chums as Buster… he gave me the exciting news—although only 17, he had been accepted into the Royal Flying Corps, and was to leave for training as soon as he was 18, only a few days off. It was the last time I was to see Alan McLeod."

The horse story would have impressed British recruiting officers. In assessing suitability, they often asked an aspiring pilot if he rode horses, apparently convinced that having "a good seat" showed athletic ability.

The first flying units of the RFC were elements of the British Army flying missions in support of the army, such as artillery spotting and reconnaissance. Many RFC pilots started out in army units, transferring to flying units from army battalions overseas. The RFC's top ace, William "Billy" Bishop, had been with the 8th Canadian Mounted Rifles, a mounted infantry, when they sailed to England with the Canadian Expeditionary Force. In July, 1915, frustrated with trench mud and a lack of action, he transferred to the RFC as an observer. Many pilots were initially seconded to the RFC from their original regiments by becoming an observer. There was no formal training for observers until 1917 and many were sent on their first sortie with only a brief introduction to the aircraft from the pilot. Originally, the observer was in command of the aircraft while the pilot just 'drove' the machine. This was found to be less effective in combat than having the pilot in charge. Observers were usually taught only enough piloting to be able to land their aircraft in case the pilot was killed or wounded.

When he was 17, McLeod had tried to enlist with the RFC, but was told to wait until he was 18. On that big day, April 20, 1917, now a strong and eager recruit, he joined up. Training successfully completed at Camp Borden, McLeod was disappointed when held back as an instructor. For a short while, he instructed young men who knew just slightly less about flying than he. Then a reprieve: he was posted to 51 Squadron at Marham, Norfolk, a night-flying unit training

Copies of this Alan McLeod oil painting were distributed to air force stations across the country at the beginning of the Second World War, to establish a feeling amongst trainees that Canada had an air war heritage. *CF Photo*

on BE-2Es. The commander there was impressed: "That kid has more spirit than all the rest of the pilots and gunners in this squadron. He's wasted messing around here in England. He ought to be over there at the Front."

Losses on the Continent didn't allow trained pilots to hang around England for long, and in November McLeod was flying FK-8s with No.2 Squadron, out of Hesdingeul, near Lille. In letters to his parents, he reassured them that his work was "perfectly safe," and that the loss rate was lower than at Camp Borden.

March 27, 1918, McLeod—still just 18 and now known as "Babe," a reference to his age and relatively unformed features—and his British observer/gunner Lt. Arthur Hammond, 23, a Military Cross attesting to his reputation as a first-class shot, were up in a lumbering "Big Ack," an Armstrong-Whitworth FK-8 bomber, after being called from bed to bomb and strafe at Bray-sur-Somme. The FK-8 was ungainly but sturdy and liked by its crews. In heavy fog that had grounded a squadron of Camels, McLeod insisted on carrying out their mission, in a battle that would see Richthofen, the Red Baron, claim three more victims. "He would take on anything," a former crew member attested of McLeod in The Royal Flying Corps in World War 1. "He would turn around to me and laugh out loud." McLeod seemed to fit the mold of ideal characteristics of the pilot as put forth by RFC commanders: "Young, unmarried, athletic, alert, cheerful, even happy-go-lucky…initiative and a sense of humour."

Hammond sat astride a rotating stool fastened to the plane's floor, and could rotate his machinegun on a half-circular gun Scarff ring. Descending on their bomb run, Hammond tapped McLeod's shoulder and indicated a Fokker tri-plane diving at them. The two had perfected a drill and McLeod, flying the aircraft as though it were a fighter, put them into a continuing turn. This made it difficult for the enemy pilot to bring his machine guns to bear on the bomber, but crack-shot Hammond could easily bring his gun around and shoot down the Fokker scout.

Eight more tri-plane scouts pounced on them. McLeod believed it was more dangerous to run, so engaged in battle. One, piloted by Lt. Hans Kirschstein, raked their bomber with gunfire, hitting Hammond twice. McLeod was hit once, in the leg, by another attacker. After recovering from a fainting spell, Hammond downed another enemy fighter. Now Kirschstein's machine gun bullets succeeded in hitting the RFC plane's fuel tank and, according to a *FlyPast Magazine* article titled Teenage Tiger, "The fuel immediately ignited and, fanned by the slipstream, blew back along the bottom of the fuselage under the cockpit seats. In no time, the flooring under McLeod and Hammond had burned away—Hammond's stool fell away, and McLeod's wicker seat was set alight,

THE LOST CANADIAN ACE

CAPT. STANLEY WALLACE ROSEVEAR was an RAF Camel and Triplane ace with 25 kills when he died in a flying accident, in 1918. For years, he was listed by the Americans as their second leading First World War flying ace (*1001 Flying Facts and Firsts*, Joe Christy), second to Eddie Rickenbacker (26 victories), but was dropped from their list when it was established that Rosevear was a Canadian. It's difficult to understand the Americans' mistake: an article by George A. Fuller in the Fall 2008 issue of CAHS refers to a 1917 action involving "Flt Sub Lt. S.W. Rosevear" of "No. 1 (RNAS) Squadron"—just before the Royal Naval Air Service and the Royal Flying Corps became the Royal Air Force. Not many of his countrymen recall Rosevear's name.

and his flying boots and long leather coat were badly scorched.

"With no cockpit floor, his seat burning and his instrument panel starting to smoulder, Alan climbed *out* of the cockpit and stood on the aircraft's left lower wing root, having first closed the throttle and attempted to set the rudder pedals to neutral—all this in what was virtually a furnace! From this precarious position, reaching through the flames to grasp the control column, he put the doomed bomber into a side-slip in order to keep the flames away from himself and Arthur."

By this time the badly-wounded Hammond was clinging atop the fuselage, feet on the bracing wires, hands clutching part of his Lewis gun. The attacking German formation broke off, but one returned to finish off the Britisher. "Despite his wounds, and the fact that his cockpit was a mass of flames and his rotating stool had fallen through the area where the cockpit floor had been, Arthur, still grasping the Scarff Ring and standing on the bracing wires, engaged the Fokker with his Lewis gun. So accurate was his fire that the scout fell away and was seen to crash in the German reserve trenches below." This action was witnessed by South African and British troops in the forward trenches, who confirmed four enemy losses.

Still standing on the wing and side-slipping to escape the flames, McLeod levelled off as they neared the battleground, with the plane engulfed in flame and choking smoke. Hammond straddled the fuselage, facing aft. Their burning plane crashed in no-man's land, ending up in a bomb crater, and under enemy fire. McLeod was thrown clear, Hammond was unconscious near the burning wreckage, partially buried in mud. There still were eight bombs and a thousand rounds of ammunition in the wreckage. "Wounded in five places, "said the *FlyPast article*, " with his flying clothing torn and burned, the teenage Canadian dragged himself over and managed to pull Arthur clear. Still under fire and unable to walk, Alan alternately rolled and carried his gunner on his back, crawling through the mud and shell holes of no-man's land. In doing so, he was wounded again, this time by a shell splinter."

McLeod finally reached the British forward trenches. He had been wounded five times, and his leg was shattered. Taken out next day, he and Hammond went through Boulogne and Dover to London. Hammond's leg had to be amputated. McLeod lay near death for many days, then recovered enough to receive the Victoria Cross, which many felt Hammond should have had as well.

McLeod's father, a Stonewall physician, hurried to his hospital bed in England, cared for him for two months until he could appear before King George V to be decorated. Before long, Second-Lieutenant Alan

Bristol Fighter, one of the most successful combat aircraft of the war, could dive faster than any other fighter.

Avro 504 on skis for Canadian conditions (left), and Royal Aircraft Factory SE-5a (right), second only to the Sopwith Camel in reputation. *Photo: Courtesy Airforce Magazine*

BRITISH "GIFT" AIRCRAFT TO CANADA

DE HAVILLAND DH-4 AND DH-9—Twenty-three, flown by Canadians in combat in both the RFC and RNAS. Used on day bombing raids and reconnaissance. In post-war period, used by CAF pilots for refresher training, aerial photography and forestry patrol. Crew: one pilot, one bomber/gunner. Max Speed, 119 mph (192 km/h). Ceiling: 16,000 ft (4,877 m), endurance: 3.2 hours. Armament: twin synchronized Vickers machine-guns forward and one Lewis machinegun aft plus, up to 450 lbs (204 kg) in light bombs.

ROYAL AIRCRAFT FACTORY SE-5A—Twelve, second only to the Sopwith Camel in reputation, flown by RFC aces such as Bishop, Ball, Mannock, and McCudden. Admired for its stable and predictable flight characteristics, equal to most other fighters in speed and agility.

BRISTOL F-2B—Two, one of the most successful two-seat combat aircraft of the war, 5,000 built. Its robust design, powerful engine, maneuvrability and relatively heavy armament enabled it to excel as a fighter aircraft. Could be dived at a higher speed than any other fighter. Numerous Canadian airmen flew Bristol Fighters, one of the most celebrated Lt.-Col. Andrew McKeever, who achieved 31 victories. One airworthy BF-2B is owned by the Canada Aviation Museum. Max Speed: 198 km/h (123 mph); three hours' endurance.

SOPWITH CAMEL—Biplane, top speed 113 mph. The aircraft had a reputation as being difficult to fly but performed great service, destroying more enemy aircraft than any other type.

AVRO 504—Biplane trainers, 62; Top speed, 85mph. Wingspan, 36 feet; ceiling 12,000 feet.

FELIXSTOWE F-3—Two-wing flying boat, 8; two 360 hp engines; wingspan 104 feet; max. speed, 87 mph.

CURTISS H-16 flying boat, 2; two 160 hp engines; biplane; wingspan 76 feet; max. speed, 102 mph.

FAIREY C-3—Two-engine biplane flying boat.

SOPWITH SNIPE, 2; two-wing; 320 hp engine, speed 156 mph. Considered one of RAF's top fighters when war ended.

McLeod returned to Winnipeg, there becoming a victim of the terrible world influenza pandemic, the "Spanish Flu," and died at age 19—five days before the November 11 armistice.

CAMP BORDEN'S CHANGING ROLE

WHEN THE BRITISH GOVERNMENT bailed out of flying training in Canada at the end of the war, it tidied things up at the airfield portion of Camp Borden by selling it to Canada. Dozens of buildings could house more than 1,500 personnel. The first aircraft ever owned by the Canadian government were warplanes. After the Air Board was formed in 1919, an "imperial gift" of more than 100 surplus land aircraft, seaplanes, kite balloons and airships was offered and, after back-and-forth discussions, sent to Camp Borden. Britain figured it could kick-start the air forces of the Commonwealth. The Brits wanted to get rid of all the now-surplus fighters, scouts, trainers and what-nots littering airfields and factories in England.

Was this an omen, the first example of Canada depending on the designs of others, rather than designing and building its own planes? Canadian military brass, who had denied the existence of aviation, were scratching their heads. How would they use the planes? What kind of air force would they finally agree upon?

At least nine aircraft never left their crates. One F-3 flying boat was assembled, registered, but not given a certificate of airworthiness and never flew. There weren't enough mechanics at Camp Borden, so Canadian Vickers in Montreal was contracted to assemble the Fairey C-3 and at least one F-3 flying boat, while two F-3s were shipped directly to Jericho Beach station, Vancouver for assembly. Other F-3s in crates were sent to Victoria Beach, Manitoba and Dartmouth, Nova Scotia. The force's three other stations were at High River, Alberta, Rockcliffe, Ontario, and Roberval, Quebec.

Camp Borden was a busy airfield in the 1920s, with thousands of flights and thousands of flying hours logged. Former wartime pilots could spend a month every two years on refresher courses. Several aircraft took part in a trans-Canada flight in 1920, a kind of silent-screen-movie haphazard public demonstration, a relay with exchanges of aircraft types and crews, and a crash or two. It began in Halifax, eventually made it to Vancouver. The same year, an Avro 504 and a Bristol F-2B flying out of Rockcliffe

Curtiss HS-2L flying boat like this was used in first bush-flying operation in Canada in 1919, at Lac-a-la-Tortue, Quebec. *Photo: Coutesy Airforce Magazine*

in the Ottawa area, were flown to see if photography in war reconnaissance could be adapted to mapping of Canadian terrain. The aircraft weren't good for the trial, but it was felt that the concept was valid. Camp Borden's 504s were used throughout Ontario in militia exercises. The de Havilland aircraft at High River, Alberta were used for forestry photography and militia work.

The air force found that the Felixstowe F-3 flying boats were not suited to use on shallow Canadian lakes and were easily damaged. Those at the Victoria Beach base, on Lake Winnipeg, were scrapped. The wartime fighter pilots had not been trained for seaplane flying, Retraining was recommended by Col. Robert Leckie, a former student of Douglas McCurdy's training school. Twelve American HS-2L flying boats were

⤞

Aircraft of the Canadian Air Force, then commanded by Lt-Col. W.A. "Billy" Bishop, buzz the flagpole at Camp Borden during the first raising of the force's "own flag" (the RAF ensign), in 1921. The ceremony was an attempt to establish the CAF's identity as a separate service from the army and the navy, and to re-enforce its links to the RAF. *Photo: Courtesy Airforce Magazine*

bought and found to need 22 inches less lake depth than the F-3s, making them more adaptable and reliable.

ADDING ROYAL TO THE CANADIAN AIR FORCE

IN 1920, THE COUNTRY'S AIR force consisted of fewer than 200 persons. Twelve years after Douglas McCurdy first advocated a national air force, Canadian Air Force officers and airmen saluted "their new flag" for the first time on November 30, 1921, after the Royal Air Force ensign had been hoisted. The ceremony was an attempt to establish the CAF's identity as a separate service from the army and the navy, and to re-enforce its links to the RAF. It was commanded by Lt-Col. W.A. "Billy" Bishop, the leading ace of the British Empire and the first Canadian aviator awarded the Victoria Cross. As it turned out, the CAF followed the marching orders of the Army.

In 1922, the department of national defence's chief of staff, Maj.-

Canadian Aces Bishop and Barker join forces post-war.

Gen. J.H. MacBrien, thought that it would be an advantage to have another Victoria Cross winner as a member of the CAF, and helped get William Barker, the most decorated Canadian war ace, a short-service commission. Barker had been a wing commander in the RFC, so he was given the army equivalent rank, lieutenant-colonel. Shortly he was sent to Camp Borden to assume command. He quickly made an impression, reportedly driving a Rolls Royce recklessly on the rough terrain, and riding a white horse to inspect the flying field. Barker had lots of ideas, such as building a fighting-capable air force, while the preferred, and legal, role was working at civil interests. He advocated an RAF-type of program, ground attack by aircraft, at long range and low level, and air reconnaissance, war-like flying that was not popular any more with the command structure. But budgets and flying hours had been cut until in the winter of 1923-24, there was no flying. Barker left Camp Borden for Ottawa to serve as acting director of the Canadian Air Force, and thus quickly ended up as the senior officer of the new RCAF. But this was not a good time for a restless airman with ideas, and a service with no funds. Barker resigned in 1926.

Not a single new military pilot had been trained at Camp Borden after November, 1918. In the summer of 1923, a training program aimed at university students was meant to bring "new blood" into the Canadian Air Force (CAF). The course ran three summer terms on the Avro 504K, and later the Sopwith Camel for advanced training. In December, 1924, the first Royal Canadian Air Force (RCAF) wings parade took place for six graduates. The RCAF took over CAF duties, but went even further and assumed responsibility for all the regulatory powers of the Air Board over civil aviation.

"The RCAF's civil functions were focused inwards, towards the economic development of Canada's hinterland, while the force's service traditions were directed outward to Britain and the Empire-Commonwealth," according to *The Creation of a National Air Force*. "Even while its primary concern was civil flying operations, the RCAF preserved a measure of military identity by participating in exchanges with the Royal Air Force…Then, when the somnolent 1920s gave way to the international anarchy of the 1930s, the RCAF was converted into an almost exclusively military force. Since it alone had the potential to defend the Atlantic and Pacific coasts from air and sea attack, it became the favoured of the three services, charged with the responsibility for providing Canada's first line of direct defence."

Maj.-Gen. MacBrien insisted it was all nonsense when he first heard of the various civil operations being carried out, but changed his mind and "swung to the other extreme, saw the opportunity of building an air force on it and gained public support for a big Air Force through the useful work they were doing", said J.S. Wilson, controller of civil aviation in a letter to a colleague in 1932. "This unusual and predominant association of the RCAF with civil duties in the 1920s and 1930s caused it to be occasionally called the Royal Comical Air Force…merely to indicate the uniqueness of its duties among the world's air forces." The term came into brief use again, after Hellyer's 1968 merger of the three services.

In 1927, Wallace Turnbull's variable-pitch propeller was tested for the first time in flight and, in 1929, Canada's first military aerobatic team, the Siskins, was formed at Borden. Throughout the inter-war years, Camp Borden was the centre of Canadian aviation, providing specialized training such as the "blind flying course" (instruments). In the mid 1930s, with 27 officers and 179 airmen, Camp Borden was home to one third of the staff of the RCAF. The camp comprised a flying training school, army co-operation school, air armament and bombing school and a technical training school. •

The Norseman (this one with the RCAF, which performed civil duties) was designed and built in Canada at the Noorduyn plant in Montreal. It operated on wheels, skis or floats and carried 10 persons. The prototype was flown in November, 1935, and many are still flying today. One was used to develop forest-fire fighting techniques. *Photo: Courtesy Airforce Magazine*

CHAPTER THREE

CIVIL AVIATION

THUNDERING, THE JET FIGHTER travelled three-quarters of the runway length, banked sharply to 90 degrees and pulled in a high-G turn so tight it remained within the perimeter of the airfield for three circuits. A brand-new "Sword" driver for 410 Fighter Squadron, I watched with fascination the most expert and brave piloting of a Sabre jet I had yet seen. We were at No.1 Fighter Wing, North Luffenham, Rutland, England. 1954.

Back at the runway, the pilot—Flight Lieutenant Dean Kelly, of brother 441 Squadron—levelled, flew dead away until he was a grey toy, turned once again, and headed back. This time, he took forever to arrive. He was slowing continually. By the time he reached the runway, the aircraft's nose was dangerously high, and it looked like it would settle. Speed brakes and flaps were out. The tailpipe was blasting the runway.

"He's going to stall," someone muttered. Everyone else's expression mirrored the same concern. Kelly cleaned up the aircraft. Throttle full on, nose gradually lowered, speed building. Suddenly, the nose lifted dramatically into a climb. We held our collective breath. Kelly took the Sabre to the top of a loop and rolled it off. It appeared to just hang. Then it powered away.

One of a phalanx of outstanding aviators flying with the RCAF, Kelly—Sabre jockey extraordinaire, squadron commander, pilot's pilot—chose on retirement to take up a line of work that would make use of all his highly-developed talents—water bombing of forest fires. He flew adapted Second World War Grumman Avenger torpedo bombers. Many others, such as "Turbo" Tarling, who logged thousands of

Early fire suppression was provided by pontooned bush planes scooping up water from nearby lakes.

hours on the air force's T-33 Silver Star ("T-Bag"), flew an assortment of Second World War leftovers, including the Douglas B-26 Invader.

Many years before these pilots joined the fire control force, two Canadian flyers employed by the Ontario government were determined to find an efficient way to fight forest fires using airplanes. When they found the answers, they added an important dimension to the store of aviation developments that civil aviation in Canada has contributed to world aviation.

This CL-215, which revolutionized fire fighting from the air, around the world, was the idea of a couple of Ontario government workers that was developed by Canadair in the 1960s, and updated with the CL-415. *Photo: Ontario Ministry of Natural Resources*

BIRTH OF THE WATER BOMBER

LIKELY THE FIRST DOCUMENTED example of an aircraft helping to suppress a forest fire occurred in the summer of 1921, in the Sioux Lookout district of Ontario, when R. N. "Reg" Johnston spotted a fire on an island. He returned to base and picked up a ranger with the necessary fire-fighting tools. The fire was extinguished before it got out of control. But the important work that led to today's effective fire-fighting airplanes, the CL-215 and the modernized CL-415, was done in the 1940s and 1950s by two pilots of the Ontario Provincial Air Service (now the Ministry of Natural Resources).

In the mid-1940s, Carl Crossley, a pilot-engineer with the service, was stationed at the Temagami base. Crossley reasoned that if military aircraft could carry large loads of bombs against enemy forces, why couldn't civilian aircraft bomb forest fires with loads of water? So, testing ideas in a high-wing Stinson Reliant, or an open-cockpit biplane, he tried taking water directly into the floats. Problem was, he had no idea how much water was entering the floats, which could mean disaster. And there was no fast way of dumping the water quickly.

Later, he was able to rig a Noorduyn Norseman that took on 100 gallons of water and dumped the load in nine seconds on a fire near Temagami, in 1945. It gave a fire crew a chance to get in and put it out.

In the 1950s, Tom Cooke, a fellow air service flyer and former RCAF Canso pilot, liked Crossley's ideas. He took them a step further, and with the aid of engineer George Gill, mounted open-top tanks on each float. "These roll tanks could be easily filled by simply moving the aircraft rapidly along the surface of the water," according to an Ontario Ministry of Natural Resources web page. "A series of cables and pulleys allowed the pilot to dump the load and the tanks, weighted at the bottom, would automatically right themselves, ready for the next pick-up. Some success at last."

Single-engine Beavers were the first airplanes to be fitted with an 80-gallon system, followed by the larger piston with bigger tanks, and then fitting of a 210-gallon belly tank.

The single-engine Otter, above, was used in early experiments with "roll tanks" to fight forest fires.

Western Canada Airways' Norseman (a mainstay of early bush flying) CF-BAU at Lac du Bonnet, Manitoba, circa 1940. Pilot Wally Carrlon stands on the float. *Photo: Courtesy Brian Johannesson*

Cooke was able to prove his theory in the summer of 1957 by dousing a fire near Sudbury. His lone Otter with roll tanks was able to hold a strip of fire about one mile long until the ground crews could get in. Without aerial water bombing, the fire would have become unmanageable.

The more powerful Turbo Beaver in 1965 revived the idea of taking water directly into the floats. In collaboration with Field Aviation, this original 1944 idea was perfected, and later adapted to all OPAS aircraft. Crossley's dream of aerial firefighting through water bombing had come full circle.

By the mid 1960s, water bombing had become a reasonably practical operation and spawned aircraft designed solely for that practice. After wide-ranging discussion of problems and recommendations in 1963 by everyone concerned with fire-fighting, Canadair (now Bombardier) had designed the CL-215 amphibian,

Western Canada Airway's "office" in the north.

which was capable of performing other duties as well. It could scoop up a full load of water within a specified distance. The first user of the CL-215 was France. Both the CL-215 and the 415 are operated in Ontario, Quebec, Manitoba and many other areas, including Corsica, Thailand, Venezuela, Germany and Italy. Most are painted in bright yellow and red, to be easily seen during low-level action. Both models carry systems that allow foam to be added to the water to improve its fire-suppressing ability.

Back in the '40s, Carl Crossley's main problem had been with air service management. "They didn't share his enthusiasm for the concept," said the Natural Resources document. "Shortly afterwards, he left the OPAS and tried to market his idea to the federal government. While they listened, little action was taken."

This reluctance on the part of governments to assist the progress of Canadian aviation was there from the time of McCurdy's first flight.

APATHY STALLED AVIATION

MAKING AN AVIATION COMPANY fly in Canada in the early days was like piloting an underpowered aircraft through torrential rain, lightning and wind shear, while flying low on fuel, and at night, without instruments. The federal government simply ignored all aviation development, civil and military. For the air-minded it was, "Heart in the sky, feet on the ground."

There was no government program to assist civil aviation in any way up to 1927, neither by direct

No flying until the engine oil is heated. Junkers CF-ATF at Eskimo Point in the Barren Lands, circa 1938. *Photo: Courtesy Brian Johannesson*

JIM SHILLIDAY | 53

VEDETTE

The Vickers Vedette was a sturdy workhorse as aviation opened up Canada's hinterlands. After many years of dedicated labor, volunteers at the Western Canada Aviation Museum, Winnipeg, restored a relic recovered from northern lake waters.

THE VEDETTE WAS THE FIRST COMMERCIAL aircraft to be built for civil aviation, by Vickers, of Montreal, which eventually would build many other aircraft types. De Havilland Aircraft of Canada Ltd. began assembling Gypsy Moths under licence in 1928 in a small Toronto plant. By 1930, it had enlarged the plant and produced 192 Moths of its own design. Also established in 1928 was The Reid Aircraft Company, at Montreal. The Fairchild Company built a large aircraft manufacturing plant in Montreal in 1929.

Canada slowly developed expertise in building bush planes. The first was the Fairchild Super 71, but since it retained the original U.S. Fairchild 71 wing, it wasn't completely Canadian. The Noorduyn Norseman, however, was completely designed and built in Canada at Montreal's Noorduyn plant. Fast, roomy, comfortable and economical, it could operate on wheels, skis or floats, and carry 10 passengers and crew. The prototype was flown in November, 1935, and many are still flying today.

James Richardson's latest in air terminals, in the late 1920s. He would not have believed the Winnipeg air terminal named after him now.

subsidy, air mail contracts nor building airports, wrote K.M. Molson in the impressively encyclopaedic *Canadian Aircraft Since 1909*, co-written with H.A. Taylor. "This was probably caused by the Government being heavily involved financially in bringing together the many railway companies that were in difficulties and the formation of Canadian National Railways." And probably they were convinced that trains had more potential than planes. Later, the railway operating procedures and organization would prove an essential model if airlines were to run economically. "Apathy and ignorance among the politicians of the day prevailed. Government was on the cusp of losing the initiative in the regulation and development of aviation," wrote Patrick Fitz Gerald in *CAHS*. Still, the aviation-minded would not be held back. In 1910, the first airplane flight in Quebec occurred, and a few days later, the first flight over a Canadian city, Montreal.

Canadian authorities would have closed down the Garden of Eden if it had been in any way connected with aviation. In 1914, Douglas McCurdy, now 28, had wangled an interview in Ottawa with Sam Hughes, Minister of Militia. He asked Hughes to consider letting him manufacture aircraft for the military. Hughes is said to have replied: "The aeroplane is an invention of the devil and will never play any part in such a serious busi-

Planes didn't fly in frigid northern conditions until the oil was thinned by heating. The first "nose hangars" were invented by Sioux Lookout brothers in the 1920s.

ness as the defence of a nation, my boy!" Perhaps old Hughes was chagrined by the cock-up with the CAS that clearly suggested he was not competent.

In 1914, McCurdy organized, and became president and general manager of the Curtiss Aeroplane and Motors Limited, a subsidiary of the Curtiss Motor Company (Baldwin had turned to hydrofoil experiments). Then he and his former experimental-flying colleague, Glenn Curtiss, decided to revive the old Baddeck-Hammondsport relationship by working from both sides of the international boundary. Curtiss turned out aircraft engines in Hammondsport, New York, and made planes by the hundreds—Model Ts of the air world—in a new Buffalo plant and shipped them to Britain.

There was a catch. The U.S. was a non-belligerent, morally bound by a national war policy to remain at peace. But would the planes be instruments of war if they were sent to Britain without vital ailerons? That satisfied the ethical conundrum, and away they went, sans ailerons (similar subterfuges were used during the Second World War to get American-made planes into Canada, and to war. Bombers were delivered to the international border, pulled across by teams of horses into Canada, and then flown to where they were needed). McCurdy now turned out ailerons in his Toronto plant to fit the Curtiss aircraft. By war's end, he had acquired all of the company stock, and ran the whole show. In a deposition, he valued the company's assets, including patents, inventions and association with Glenn Curtiss at a sizeable, for the time, $7 million.

Ellwood Wilson, a forester, set up the first bush-flying operation in Canada in 1919. He stationed two war-surplus Curtiss HS-2L flying boats at Lac-a-la-Tortue, Quebec. In 1922, Wilson founded Fairchild Aerial Surveys Co. (of Canada), the first aviation company in the country to operate year-round. Civil aviation activity picked up when Wilson's little bush company became Laurentide Air Service Ltd., of Montreal, in 1922, and expanded after getting lots of work from the Ontario government. It operated cargo and mail service into Quebec's goldfields in what

A postage stamp honoured Alexander Graham Bell's contribution to Canadian aviation

Lockheed Vega of Canadian Airways, in 1930 Canada's first commercial airline. *Photo: Courtesy Airforce Magazine*

STEVENSON FIELD

BOTH JAMES A. RICHARDSON AND WINNIPEG, the spawning ground of Canadian aviation, were suitably recognized in 2007 when federal Justice Minister Vic Toews announced the city's airport would be renamed James Armstrong Richardson International Airport to honour the man who established the country's first commercial airline.

The airfield had been named Stevenson Field in 1928 after Frederick J. "Stevie" Stevenson, war ace and aviation pioneer. Stevenson grew up and was educated in Winnipeg. At 18 he went to war, was with the Royal Flying Corps by 1917. In France, he destroyed 17 enemy aircraft and three observation balloons. At war's end, he was a captain holding the Distinguished Flying Cross and France's Croix de Guerre. He ferried peace conference diplomats between London and Paris, then served as an RAF flying instructor in Russia.

In 1920, Stevenson joined Winnipeg's Canadian Aircraft Company, flying widely in Manitoba and Saskatchewan, contracting with towns for aerobatic exhibitions and short flights for passengers. He switched to Richardson's Western Canada Airways Ltd., flying heavy equipment in open-cockpit Fokker Universal aircraft from a base along the proposed Hudson Bay railway route, in severe winter conditions. These flights aided in the selection of Churchill as an ocean terminus, and marked the beginning of large-scale freighting by air in Canada. For his flying feats, he was the first Canadian awarded the international Harmon Trophy, as Canada's outstanding pilot, in 1927.

Stevenson died in a 1928 crash during a test flight out of The Pas. He is buried near Winnipeg's airport, in Brookside Cemetery. Months later, at the airport opening, a plaque was unveiled: "This aerodrome is named Stevenson Field in dedication to the late Captain F.J. Stevenson of Winnipeg, Canada's Premier Commercial Pilot." When the airport was renamed Winnipeg International Airport in 1958, the plaque honouring Stevenson still was on display, as it probably will continue to be in the city's new terminal.

Fred Stevenson was a force in developing large-scale freighting by air in Canada, and had a city airport named after him.

became the Rouyn-Noranda district. But in 1924, Ontario set up its own air service, and Laurentide collapsed the following year.

After the war, droves of Canadian pilots came home to a country where authorities ignored aviation. So anyone with goggles and a small plane scrabbled and scratched the countryside for whatever was available—barnstorming and joy-riding, once-in-a-while charters, the odd photographing assignment. But from 1923 to 1926 hardly a pilot was flying.

Foreign-designed aircraft skis for Canadian bush flying had proven inadequate for the rough landings, and now a Sioux Lookout, Ontario firm, Elliott Brothers, designed and built hardier skis that would be popular for many years, even being used on three Antarctic expeditions. This company also came up with an all-weather canvas "nose hangar" to service aircraft engines outdoors, a great help in bush-flying operations.

(Right) A "Flying Jennie", owned by war ace Wop May, who tried to make an aviation business "fly" in Edmonton. *Photo: Courtesy Airforce Magazine*

GRAIN BARON AND AVIATION PIONEER

THE AIR BOARD THAT HAD BEEN set up in 1919 had regulatory powers that allowed it to use government-owned planes for civil operations—a proviso that persuaded the anti-air force group in government to go along with a flying component in the military. HS-2L flying boats from an air station in Dartmouth, Nova Scotia, were used to patrol vast Quebec timber areas, and were said to have been the first "bush planes." Then, Canadian civil aviation slowly established itself in 1926 and 1927 when planes finally were seen as the best way to rush workers and supplies into a rich gold strike in Ontario's Red Lake district. Bush-flying entrenched itself when Red Lake attracted Western Canada Airways Ltd., founded in 1926 by James. A. Richardson, of Winnipeg, who pretty well carried the aviation business in Canada until the start of the Second World War.

Richardson was a grain exporter during the rich years of Canada's wheat economy. He was an entrepre-

GOVERNMENT DOES SOMETHING RIGHT

ALTHOUGH DOUGLAS MCCURDY may have pushed for civil air regulations and establishment of an air force earlier, it's fair to believe that John A. Wilson had a huge claim to the accolade, "Father of the Royal Canadian Air Force." Wilson worked from inside the government, which gave him the advantage for getting results, while federal officials may have regarded McCurdy simply as a pest.

Wilson, an engineer, came from Scotland. After some years, he was deputy minister of the naval service department. He submitted a paper, at the end of the First World War, titled Notes on the Future Development of the Air Service in Canada Along Lines Other Than Those of Defence. It asked Ottawa to devise a national air policy covering air regulations and flying operations. First, civil aviation should be developed, followed by an air force. Politicians were as uninterested in aviation as before, but now, according to esteemed Canadian military historian and author Hugh A. Halliday, in *Legion Magazine*, "several groups, including forestry associations, were urging adaptation of aircraft to civil ends, while surveyors and mining officials inside the federal bureaucracy joined Wilson…in raising the same concerns. Consequently, Wilson (had been) assigned the task of writing the Air Board Act", which became law in 1919.

"The initiative for responding to the challenge of the air age was left to a small group of middle-ranking civil servants…like J.A. Wilson…in converting the expansive potential of aviation, …to constructive peacetime uses, a focus of commitment which coincided nicely with the prejudices of an unmilitary people tired of war and a government bent on economy," wrote W.A.B. Douglas in The Creation of a National Air Force. "The real foundations of a nation's airpower, it was reasoned, lay in the widespread development of civil aviation, including extensive commercial operations and a healthy aircraft manufacturing industry. This in turn would provide the foundation on which a military air force might later be built."

From 1923 on, Wilson wore many government hats, including secretary of the Air Board, assistant director and secretary of the RCAF, controller of civil aviation (when he became a Trans-Canada Airlines director), and director of air services. He was often called The Father of Canadian Civil Aviation. "…but in a very real sense," wrote Halliday, "he also ensured that between the world wars the RCAF would form, survive, and prosper, albeit as an organization as much dedicated to civil flying as to military preparations."

In 1939, Wilson took over the job of involving civil aviation in the country's war effort. Then his department was made responsible for developing sites for training stations when the Commonwealth Air Training Plan was adopted. The flying clubs that Wilson had championed in 1928 set up many of the flying schools. He helped form ferry command, vital to the supply of warplanes to Allies in Europe, and in demonstrating the practicality and safety of trans-Atlantic flying for passenger services. He was involved in establishing the International Civil Aviation Association (ICAO) in 1944, to be headquartered in Montreal. In 1973, Wilson's name was added to the Canadian Aviation Hall of Fame.

The air force sponsored Canadian flying clubs, and each was issued two training aircraft from RCAF stocks, followed by technical advice. The clubs in turn became the nuclei of auxiliary air force squadrons after 1934. The RCAF undertook erection and equipping of airfields for airmail service in 1927, after Wilson's recommendation for the service was adopted.

John A. Wilson, above

(Above) A de Havilland Beaver. (Inset) Western Canada Airways air freighter unloading supplies.

THE BEAVER, A BUSH-FLYING MARVEL

IN 1943, CANADIAN AIRWAYS and 10 bush-flying firms joined together to form country-spanning Canadian Pacific Airlines and continued to service the northern areas. A bush-flying marvel also was born in that decade—the de Havilland Beaver. The company consulted with bush pilots and the Ontario flying service and produced a STOL (short take-off and landing) aircraft—referred to as a half-ton flying pick-up truck—that had no equal, and more were sold (almost 2,000) than any other Canadian-designed and built aircraft, to this day. It caught on around the world after the U.S. Army bought them for service in the Korean War, in which they served as "flying Jeeps." They were so reliable and adaptable that there is a thriving Canadian business today in reconditioning and rescuing old Beavers, and putting them back upstairs.

Bush planes were the only connection between San Antonio gold mine in Bissett, Manitoba, and civilization.

neur with the imagination to see what the country needed. He was also willing to take chances and invest, and he liked to make money—all good qualities for a young man in a young country with a future. Richardson set up an air base in Lac du Bonnet, Manitoba, to service northern mining areas, up to then accessible only by canoe or dog team. Western Canada Airways became Canadian Airways in 1930—Canada's first national airline—which eventually transmogrified into Canadian Pacific Airlines which was then folded in to Air Canada. Western Canada Airways pilots included names headed for a measure of fame in military and civil aviation, such as Roy Brown (not the Roy Brown who downed the Red Baron) and Punch Dickins, the first commercial pilot to cross the Arctic Circle. Richardson landed a federal contract to carry mail between Winnipeg and Calgary.

In 1926, the U.S. postal department was pioneering airmail services through private companies. The Canadian government, at last provided with an example, announced it would allow airmail routes in 1927. Sir Alan Cobham, knighted for his many contributions to British

JIM SHILLIDAY | 61

 Big men in the budding aviation business: James A. Richardson, right, and Ben Smith, J. G. McDairmid, and Mitchell Hepburn.

aviation (his family company would become famous for pioneering air-to-air refuelling), lectured in Ottawa on the varied uses of aviation. Canada's eccentric Liberal prime minister, W.L. Mackenzie King—who up to then had ignored aviation and abhorred the military—took to the man, arranged lunch together, and listened intently as Cobham advised that governments should assist the development of air transport by private companies and promote the establishment of flying clubs. These ideas had been put forward before by Canadians but nothing had been done. Now the prime minister was impressed.

As well, when Charles Lindbergh flew the Atlantic in May, 1927, King was enchanted by him. Lindbergh, he wrote in his diary, "was like a young god who had appeared from the skies in human form" and he wondered if they might be related. He invited the American flying sensation to Ottawa, and Lindbergh showed up in the *Spirit of St. Louis* to help celebrate Canada's Diamond Jubilee in July, and stayed at Laurier House. King saw how dangerous Lindbergh's passion could be when, several weeks after the Ottawa visit, one of the 12 U.S. Army planes escorting Lindbergh crashed on landing, killing the pilot. It's likely that the Canadian Government's change in attitude towards civil aviation was due to King's interest in Cobham's ideas, and his worship of Lindbergh.

The Canadian Post Office did start letting air mail contracts in 1927 to inaccessible areas such as the north shore of the Gulf of St. Lawrence, and then a main route from Halifax to Windsor/Detroit, the country's most populous area. A route through the Prairie Provinces followed, from Winnipeg to Edmonton and Calgary. The Great Depression disrupted plans for connecting eastern and western routes. The start of airmail routes was of little benefit to the Canadian aircraft manufacturing industry. In fact, Douglas McCurdy's Curtiss Reid company had designed and built the Courier specifically as a mail plane, a high-wing monoplane that could be fitted with floats. It was smaller, as an economy measure, and easy to fly, but had been introduced at the wrong time.

James Richardson's Western Canada Airways was the lone big company in the country, operating from western Ontario westward into British Columbia and north to the Arctic Ocean. There were a few small companies in the east, but WCA was the best organized and financed. When Richardson was told Americans were planning to take over the eastern companies, he struck first, setting up a group that bought them all. Control of Canadian air transport should remain in Canadian hands, he insisted. The companies soon were joined to form Canadian Airways Ltd., operating from coast to coast, that took to the air on November 25, 1930. CAL was bought by the CPR in 1941.

The Air Board had been scuttled in 1923, and the department of national defence took over until 1936, when a new department of transport was formed and assumed command. It was an uncertain period for civil air regulation, with a climate of, "Dog eat dog and the devil take the hindmost."

Canadian Airways almost had to close down in 1932 when the economic depression ended airmail services between all the big Canadian cities. The company sank a lot of money into buying aircraft and buying or leasing facilities for airmail. Now it had no revenue to pay what it owed. But gold again came to the rescue. Because of high prices, prospecting and exploration activity continued. Mining kept civil aviation from spinning in, and more freight was being moved in Canadian skies by the mid-1930s than in the rest of the world in total—without a cent of government money to help bush flying operations. But in the whole country, only Moncton, New Brunswick and Charlottetown, Prince Edward Island, were linked by air travel. It's difficult to imagine any other developed country having such poor air service between major centers.

Politicians have always been adept at whispering suggestions to businesses, especially like-minded businesses, and when mail services were cut in 1932, Canadian Airways was given the impression that when the Depression ended, a trans-Canada airline would be set up and CA would operate it. So the airline hung on by its wingtips. CA and govern-

ment officials planned together for a 1937 opening of a new country-wide airline. As can happen so easily in that kind of milieu, things changed. C.D. Howe, a former American who was now so powerful in the Liberal cabinet that he was referred to as "the minister of everything," announced that the government would set up its own Trans-Canada Airlines, mostly run by hired Americans. In 1937, TCA ran the short Vancouver-Seattle route, then, in July, one of its two-engine Lockheed Electras—with Howe on board for a publicity-grabbing test run—flew 4,025 km from Montreal to Vancouver, after stops at Kapuskasing, Sioux Lookout, Winnipeg, Regina and Lethbridge. By 1938, TCA had expanded daily service east to Montreal and, by 1939, to Moncton.

Forty-six years after the *Silver Dart*'s wake-up call for Canadians to get with the Air Age, and after years of civil aviation confusion across the country, the Air Transport Board finally was set up to regulate civil aviation in matters of competition, allocation of routes and areas of operation, and control of fares. •

Present-day view from Crow Hill looking west along the first airfield at Harbour Grace designed just for attempts to cross the Atlantic. Amelia Earhart and dozens of intrepid aviators dared the elements in relatively flimsy aircraft, some losing the dare and their lives.

CHAPTER FOUR

COMING OF AGE:
TAMING THE ATLANTIC

NEWFOUNDLAND, IT SEEMS, was made for the dawning age of aviation. Canada's youngest province had been England's oldest colony. Just six kilometres southeast of St. John's Harbour is Cape Spear, the island's most easterly point. Stand there facing seawards and all of North America is behind you. That's why cross-Atlantic navigators, from the Vikings to incoming jetliners diverted from New York's terrorist catastrophe in 2001, landed in Newfoundland.

Newfoundland wears a decorative brooch so representative of the aviation mystique that the landing strip at Harbour Grace should be mecca to all air-minded Canadians. During the 1930s period of the RCAF's relative stagnation, this nautical location took wing as aviators pushed the envelope and helped pave the way for Canada's civil and military aviation to soar to new heights in passenger service and defence of the free world.

Several flights from Harbour Grace were wildly celebrated in their time, but the take-off and crossing of the Atlantic, west to east, by Amelia Earhart—first woman to make the crossing solo—is the one generally remembered today. For those with wings, there was no talk of inequality: women were right up there with men in dominating the sensational headlines. It was common, though, to dub these women with the now quaint-sounding designation "aviatrix."

Because of its strategic location—closest "new world" land to the European continent—Newfoundland was global flying's launching pad into the future and remains a vital link today. The province's aviation history, before it joined the Canadian confederation, is a marquee lit with flying stars—Alcock and Brown, Earhart, Rickenbacker, the Lindberghs, Post,

JIM SHILLIDAY | 65

Boyd, Balbo. We're lucky, for it is now Canadian history.

Twenty record-seeking flights were attempted by North Americans and Europeans from Harbour Grace between 1927 and 1936. Eleven succeeded, four flew into the distance and were never seen again. Two crashed on take-off, two changed their minds and came back, and one made it as far as the coast of Ireland and crash-landed. Those who disappeared likely were the victims of aircraft, engines and instruments that were crude, and their lack of training in instrument or "blind" flying.

When I found the route to the airport, I followed it until it petered out. Then I was at Earhart Road, a path of gravel heading upward to a wire fence at the foot of a 60-feet-high granite outcrop, known locally as Crow Hill. I stepped through a makeshift gate into another era: the rocky pile anchors one end of a 4,000 feet-long by 200 feet-wide runway. Here early aircraft flown by stout-hearted pilots had trundled over a flinty surface, severed earth's ties and gone after the honour of doing something no-one had done before. It was a time when flight still evoked wonder. The aura of the place was so gripping after my first visit, I flew to Newfoundland soon again, to walk the runway another time—as I had walked the path up to *Beinn Bhreagh* near Baddeck—and marvelled that under my feet was a take-off run over which had trundelled the plane wheels of great adventuring pioneers such as Amelia Earhart.

The little runway at Harbour Grace, except for grass now growing between the stones, is exactly as it was when built in 1927 to take advantage of a growing, worshipful surge to set records in the skies, to be first to fly across the Atlantic Ocean—and farther. No other 7.2 hectares in the world played such a concentrated role in the birthing of the world's commercial aviation. Modest as it seems today, in the late 1920s some active aviators swore that it was the best airfield in North America—in spite of the high rocky hill at one end that few airport designers or pilots in the world would tolerate now.

I stood upon the rocky upthrust, gazing down the unsophisticated runway into the west, and into that romantic period early in the last century when faltering experimental flying had captured the world's imagination. I stepped forward, crunching, following the path of Amelia Earhart's bright red Lockheed 5B Vega—and so many others—into aviation history.

Amelia Earhart's red-and-gold Lockheed Vega on the take-off roll for her historic solo crossing of the Atlantic Ocean. *Photo: Centre for Newfoundland Studies Archives, Memorial University*

THE LITTLE RUNWAY THAT COULD

THE PUBLIC'S INTEREST IN NEW-FANGLED flying was whetted during the First World War when the exploits of Canadian air war heroes such as Bishop, Barker and Brown were publicized. Around the world, fascination rose almost to the status of religion by the late 1920s and early 1930s. The Atlantic Ocean was the big challenge. *London Daily Mail* owner Lord Northcliffe, distracting attention from the horrors of the just-finished war, offered a 10,000-pound-sterling prize for the first non-stop air crossing in a "heavier than air machine," won in 1919 by John Alcock and Arthur W. Brown. The pair took off from primitive terrain known as Lester's field—named after the farmer-owner—just north of St. John's, in a Vickers *Vimy* carrying the first trans-Atlantic mail shipment, and crash-landed in a County Galway bog in Ireland, 16 hours and 57 minutes later. Both were knighted.

A U.S. effort in this contest was rightly ignored: the U.S. Navy sought glory by entering a team of three huge "Nancies" NC-TA flying boats powered by four 400-hp Liberty engines and products of the Glenn Curtiss factories. But the Americans weren't playing the game—they couldn't make it non-stop, so planned a longer, multi-leg, more southern flight path, with a re-fuelling stop in the Azores, and 25 U.S. Navy warships spaced all along the way providing navigation aids and a promise of rescue, if needed. The team was led by John Towers, a former Curtiss flying student. One of the aircraft, piloted by Lt.-Cdr. Albert Read, made the Atlantic crossing from water a few kilometres south of Harbour Grace, at Trepassey, the first to do so, but not by the Northcliffe rules. The achievement soon was forgotten by all but the Americans.

Harbour Gracians caught the aviation bug just before this triumph. While Alcock and Brown planned their record attempt just a few miles away, British Rear Admiral Sir Mark Kerr wheeled 105 crates containing the world's largest biplane bomber, the Handley Page *Atlantic*, into the fishing village and began assembly. But with the Alcock-Brown success, that adventure was abandoned.

Alcock and Brown's Vickers Vimy is prepped at Lester's farm field near St. John's, Newfoundland, the day before making the world's first non-stop trans-Atlantic flight, June 14, 1919. *Photo: Centre for Newfoundland Studies Archives, Memorial University*

⤻

Wiley Post (wearing white coveralls) stands near his Lockheed Vega, *Winnie Mae*, before taking off from Harbour Grace on his first round-the-world attempt in 1931. Spectators gawk from Crow Hill, at the runway's east end. *Photo: Centre for Newfoundland Studies Archives, Memorial University*

The United States sucked air in the aviation field after the Great War, cutting military funds for new aircraft designs and closing many airfields. There was no pick-up in general aviation progress until 1926, and the U.S. ignored passenger service until the late 1920s. But just three months after the war, Germany started the world's first passenger airline company, within its own boundaries; Britain and France began ferrying passengers by air in 1919 in modified bombers, between London and Paris. They were setting the scene for what would come when the Atlantic Ocean was conquered.

Then, the take-off the fledgling aviation world had been hoping for. On May 21, 1927, young Charles A. Lindbergh flew solo in *The Spirit of St. Louis* non-stop from New York (passing over St. John's to alert the world of his path and timing) to Paris. He won world adulation and a $25,000 prize—and newspapers variously referred to him as "Lucky Lindy," and "The Flying Fool."

The move to set up airline operations was given impetus. Public interest in air travel bloomed and aviation stocks skyrocketed. There was no shortage of adventurers who wanted to push the envelope of the time—in the year following Lindbergh's success, 31 Atlantic crossings were attempted, from both North America and from Europe. Ten succeeded, but 20 men and women died in the attempt.

Lindbergh's flight galvanized action at Harbour Grace. The same year of Lindbergh's triumph, The Stinson Aircraft Corporation, of Detroit, sent Fred Koehler to scout the area for a likely launching place for an around-the-world flight. He met a local, John L. Oke, who led him up to a flat plateau northwest of town between the harbour and a lake to the north. Its elevation meant there were no surrounding obstructions, except for a high outcropping of rock at the east end.

The coastal residents, eager to restore local pride because their great sealing and fishing enterprises were faltering, formed the non-profit Harbour Grace Airport Trust Company. The firm supporting the global flight added funds. The Newfoundland government granted money to clear and level the airstrip. Citizens did most of the rock and bush clearing with horses and carts, and levelled with rakes and shovels, wrote Bill Bowman in *The Challenge of the Atlantic*. They began work August 8 and finished in just under a month. Thus was born North America's first civilian airport designed for trans-Atlantic flights.

Miss Dorothy, the orange-and-green Bellanca monoplane piloted by the world-famous Jim Mollison, made the last of the pioneer trans-Atlantic flights from Harbour Grace on Oct. 28, 1936. *Photo: Centre for Newfoundland Studies Archives, Memorial University*

Workmen still were fussing with the runway's surface when, on August 26, 1927, they looked up to see the *Pride of Detroit* winging in. The high-wing monoplane, similar to Lindbergh's, would try to set an around-the-world record, the reason for Fred Koehler's first visit to Harbour Grace. Edward Schlee, president of the Waco Oil Company, and pilot William S. Brock, said the airfield was one of the finest they had seen. They spent the night, took off next morning and landed at Croydon, England, 32 hours later. For technical reasons, the attempt was later abandoned in Japan.

In September, Capt. Gerry Tulley and Lt. James Metcalfe tried to fly from London, Ontario, to London, England, via Harbour Grace, where they landed September 5. Their Canadian monoplane, *Sir John Carling*, took off next morning. Onlookers by the airstrip were the last to see them alive. Three planes trying the big hop that week were lost.

On September 6, Canadian Duke Schiller and American Philip Wood landed at Harbour Grace Airport. They had left Windsor, Ontario, in the Stinson *Royal Windsor*, headed across the Atlantic to Windsor, England. But there had been an outcry that the unusually stormy weather over the ocean had caused too many deaths, and the organizers cancelled their flight. That same year, 1927, enough people were dying in the attempt to master the Atlantic on wings that Canada's prime minister,

JIM SHILLIDAY | 69

CANADA'S LITTLE LINDY

BEFORE ANYONE ELSE HAD DONE IT, Erroll Boyd knew he wanted to fly the Atlantic Ocean. Toronto's Boyd became the first Canadian to fly the Atlantic, in 1930, and he succeeded because he had perfected better "blind flying" skills than many of the other brave-hearts taking off from Harbour Grace airfield perched on the eastern shores of Newfoundland.

Four years after the epic solo flight of Charles Lindbergh, Boyd was hailed as "Canada's Lindy." As early as 1916, he had talked of flying the Atlantic. Just six years after Douglas McCurdy made the first airplane flight in Canada, 24-year-old Boyd wanted to get up there too. Like most young Canadians, he had been gripped by the flying mystique and now, with the First World War well launched, he saw that flying and war offered the tantalizing lure of danger and adventure.

His family understood, and in 1915 paid his passage to England to join the Royal Flying Corps. But he was found colour-blind. So he tried the Royal Naval Air Service, and got in. His early flying training came from John Alcock, who—along with Arthur W. Brown—would be knighted for making the first Atlantic flight ever, in 1919.

Boyd's first combat flying had him attacking German Zeppelins, wrote Herb Kugel in LOGBOOK. This involved night flying that accustomed him to instrument flying, a vital skill that would help him so much later on. After being posted to a fighter base at Dunkirk, France, his aircraft was hit by anti-aircraft fire, he forced landed in Holland and was interned, then allowed to go to the United States, where he test flew Curtiss JN-4s. Returning to the RNAS, he was promoted to captain after it merged with the RFC, just before the war ended.

By the time he decided to attempt an Atlantic flight, he had run a Toronto car rental (there had been no career in aviation), wrote songs (his Dreams became a Broadway hit in 1924). After Lindbergh's flight, aviation picked up again and Boyd flew mail in Quebec, went to Mexico and gained more instrument-flying experience. His flying in Mexico, and a 1930 non-stop round trip between Bermuda and New York, prepared him admirably for a try at crossing the Atlantic Ocean. He now had excellent "blind flying" skills and a logbook total of almost 7,000 flying hours.

Relying on two magnetic compasses and his artificial horizon—and fellow airman Harry Connor, who kept him awake—Boyd flew low-level through the dark night and rough weather for more than 10 hours. Then a fuel distribution problem arose and Boyd had to land on a Scilly Islands beach just a few miles off the English coast. In just over 36 hours, he had flown more than 3,700 miles. He and his plane (the former Columbia, which had flown the Atlantic earlier) now renamed Maple Leaf, were famous. Boyd remained active in aviation until he died in 1960, at 69.

Toronto's Erroll Boyd, first Canadian to fly the Atlantic at a time when risk was more a part of their daily lives than now, had more instrument-flying experience that most of the "big pond jumpers".

Mackenzie King, warned he was considering legislation to halt such oceanic flights originating in Canada. He believed Canadians were against such risks. He mis-read the tea leaves completely.

Zeal easily turns to foolishness. Fred Koehler, who had picked the site for Harbour Grace Airfield and Mrs. Frances Grayson, a pilot and a navigator, took off from New York late afternoon of December 23. There was rivalry for the honour of carrying the first woman to cross the Atlantic. Ignoring weather reports, they flew overnight in Dawn towards the Conception Bay town. Early next morning, their engine was heard over nearby villages—Heart's Content, Pouch Cove, Brigus, and others that were engulfed in heavy blowing snow. They couldn't locate the airfield and ran out of fuel. No trace was ever found.

Ross Smyth wrote in *The Lindbergh of Canada: The Erroll Boyd Story*, that up to 1937, 90 airplanes had tried to cross the ocean, just 13 reached their destinations, 30 got across then faltered, and 41 flyers died.

The following June, two more women answered the call of adventure. Amelia Earhart, an accomplished flyer, landed as a passenger in *Friendship*, a tri-motor "hydroplane" or seaplane. It landed in the harbour at Trepassey, a fishing village 75 miles south of St. John's. A week later, the high-wing Bellanca monoplane *Columbia*, which flew the Atlantic from the U.S. two weeks after Lindbergh, settled onto Harbour Grace Airfield carrying a two-man crew and Mabel Boll, hoping to beat Earhart as the first woman to fly the Atlantic.

Earhart was fully aware of the dangers she faced. She sometimes left letters to be read if she

Intrepid (sometimes foolish) aviators taking off from Harbour Grace would do a U-turn onto a due east heading and find this view of the town and, as they climbed—and the fear threshold climbed with them—saw far ahead the daunting Atlantic and their fate, not always a good one.

were killed. She called them "popping off" letters. One was addressed to her father, dated May 20, 1928: "Dearest Dad: Hooray for the last grand adventure! I wish I had won, but it was worth while anyway. You know that. I have no faith we'll meet anywhere again, but I wish we might. Anyway, good-by and good luck to you. Affectionately, your doter, Mill.

June 17, after spending 13 days at the village on the Avalon Peninsula waiting for agreeable weather conditions, Earhart's group took off

 (Above) Avoiding this rocky hill (foreground), adventuresome young flyers throttling back to land straight ahead had this view of the airfield and Lady Lake, to the right. Below, to the left, stretches the town. (Inset) Amelia Earhart.

in *Friendship*, a Fokker, the first seaplane to fly the Atlantic non-stop. They flew "blind" for 2,246 miles, at an average speed of 113 mph. Earhart's plane landed in Wales, and Boll was stricken to learn next day that she would not be the first woman to cross the big ocean. Backers told the three to return to New York.

The German rigid airship, or dirigible, Graf Zeppelin's record of 21 days for circling the earth, from August 8-29, 1929, was the target when Wiley Post left New York in his Lockheed Vega *Winnie Mae*, a high-wing monoplane with a powerful engine and the latest instruments. The goal of Post and his navigator, Harold Gatty, was 10 days. Just before 11 a.m., June 23, 1931, they landed in Harbour Grace, ate in town, and took off again within four hours. They crossed the Atlantic to England in 17 hours, 17 minutes, continued to Berlin, Moscow, Siberia, Alaska, Edmonton and New York, after an epic flight of 16,000 miles (25,748 k), 13 stops and eight days, 16 hours. The city welcomed them with ticker tape and cheering throngs, their flight regarded as second in importance only to Lindbergh's in the advancement of world flight. Post and the famed actor and humorist, Will Rogers, would die together in a crash in Alaska, in 1935.

Two Hungarians took off from New York's Roosevelt Field and landed at Harbour Grace airstrip on July 13, 1931, in a racy, low-wing monoplane, *Justice for Hungary*. Capt. George Enders and Lt. Alexander Magyar waited for clear weather and went on their way two days later, heading for Budapest, 3,288 miles away. Twenty-six hours later, after flying through a severe Atlantic storm, short of fuel, they forced landed 12 miles short of their destination. More than 100,000 proud Hungarians greeted them in Budapest.

Flying in darkness and bad weather from New Jersey, Lou Reichers located the field and landed at Harbour Grace at 6:24 a.m., May 13, 1932. Four hours later, in his speedy, low-wing monoplane, *Liberty*, he was airborne bound for Paris. He went into the water near Ireland's coast, was rescued by the USS *President Roosevelt*.

Not satisfied with being "baggage" on her 1928 trip across the ocean, Amelia Earhart showed up in Harbour Grace at 2 p.m., May 20, 1932. Her plane, a high-performance wood-covered Lockheed Vega, that she had been flying for three years, was the same make flown by Wiley Post for his around-the-world flight (Post's plane had fuel tanks from Earhart's which she had swapped for the wheels Wiley had replaced with pontoons). She now had more than a thousand flying hours, much of it in "blind flight," that is, on instruments. She had added instruments, including a drift indicator and three compasses—an aperiodic, a magnetic, and a directional gyro for checking one against the other. A loner at heart, many had already insisted Earhart looked much like Lindbergh, tall, with the same long face and strong chin, the same light hair and blue eyes. She had been referred to as "Lady Lindy."

Earhart rested at Archibald's Hotel (now Hotel Harbour Grace, which serves wonderful chowder). Then she climbed into the gold-trimmed, red high-wing monoplane carrying a can of tomato juice with a straw, a thermos of soup, maps, a couple of scarves, comb, toothbrush. She waved to the cheering crowd and her two-man maintenance crew, and then flew into the sun setting over the Atlantic at 7:13 p.m. That night she endured losing her altimeter, severe icing which put her into a spin from which she recovered just above the waves, and trailing flames from a broken manifold ring weld. After 2,026 miles, she came down on the Gallagher farm, six miles from Londonderry, Ireland, 14 hours and 56 minutes after take-off, the first woman (third person) to fly solo across that ocean, and remains the only woman to do it, west to east. The Lindberghs cabled congratulations. King George sent a message. France gave her the Knight's Cross of the Order of the Legion of Honour, and United States of America, the Congressional Medal of Honour.

First World War American flying ace Eddie Rickenbacker, head of Eastern Airlines, landed at Harbour Grace in 1936 with fuel and spare parts for a Vultee monoplane, *Lady Peace,* carrying two men forced down about 150 miles away. On September 15, he landed a DC-2 transport plane, the largest aircraft seen in Newfoundland (a similar plane, a DC-3, *The Spirit of Harbour Grace*, sits in a place of honour looking out over Conception Bay today). It was an exciting week for Harbour Gracians who saw more planes on their airstrip than ever before, including three press planes from New York and Boston.

The same year, October 29, Englishman James Mollison in his orange and green Bellanca M*iss Dorothy* flew directly to Croydon Airport, London, England, the last time the airport was used for such a venture. He had made the first solo flight attempt west to North America in 1932. Mollison was the husband of Amy Johnson, whose record flight from England to Australia made her famous (she would die during the Second World War when the warplane she was ferrying crashed in the Thames Estuary). He advised the English pilot, Beryl Markham, the first woman to fly the Atlantic solo the "hard way," from East to West against the winds, and one of the last of the great aviation chance-takers of the time. When Markham took off in 1936 from Abingdon, England, in her Vega Gull, *The Messenger,* Mollison wished her luck, then said, grimly flippant to a friend, "Well, that's the last we'll see of Beryl." After 22 hours, she came down, while hoping to make Harbour Grace, in a field near the tiny community of Baleine Cove, on the eastern tip of Cape Breton.

Markham explained her feelings about the ocean flight: "I have become fidgety. It is natural. I am still young and while I am supremely confident I am not particularly anxious to die. But if I get across… it will have been worth it because I believe in the future of an Atlantic air service. I planned this flight because I wanted to be in that air service at the beginning. If I get across I think I shall have earned my place. Don't you?"

During the Second World War, the airfield was closed by the military. Huge boulders were rolled onto the runway. It based a High Frequency Direction Finding (Huffy Duffy) unit, on guard for submarines. In 1977, it would be restored to usable condition by the Harbour Grace Historical Society. Its years of being considered abandoned were ended in 1999 by reinstatement to official interna-

The Italian Savoia-Marchetti SM.55X flying boats were beautiful and unusual, double-hulled, with one engine facing forward, a "tractor", and another backward, a "pusher". *Photo: Centre for Newfoundland Studies Archives, Memorial University*

tional status, with the designator CHG2. Aside from autumn air cadet gliding activity, most action on the runway now comes from former pilots flying remote controlled model aircraft showing off their Immelmann turns, loops and wing-overs.

It's visited comparatively rarely by aircraft, its time passed by, a quiet snapshot of the past on a low ridge poking up between Conception Bay and Lady Lake.

BALBO DRIVE

ON JULY 26, 1933, the largest gaggle of aircraft to make a trans-Atlantic flight arrived at Shoal Harbour, Clarenville, in central Newfoundland, on the last leg of its spectacular flight. Benito Mussolini, the Italian dictator, had decided to set the splendour of an Italian flying-boat "armada" before the world. There was a world's fair on in Chicago, and Gen. Italo Balbo, the dictator's flamboyant Minister of Air came up with the idea of leading a mass formation of 25 Savoia-Marchetti SM.55X flying-boats from Italy to Chicago, stopping in Newfoundland on the way. A large crowd watched as the two dozen planes landed, and Italian general Italo Balbo and his crew received a hearty Newfoundland welcome. They stayed several days, and captivated their hosts.

The shoreline drive was crowded with townsfolk—and many who had travelled quite a distance—for the early-morning take-off, the air cool and still unaffected by the slanting sunlight, first one motor snorting, another, many joining the chorus. Now they were moving, swinging, positioning in the wide expanse of the harbour, the water high on their hulls that moved sluggishly. Then, by twos and threes the water birds were

Italo Balbo

Chicago, and captured the attention of the world. The Air Armada episode also was intended to celebrate the tenth anniversary of the Italian air force, Regia Aeronautica.

On his return to Rome, General Balbo was elevated to the rank of Air Marshal by Premier Mussolini. Then came war, which Italy entered on the side of Germany. Balbo was against the war and it was said that, since Balbo's popularity rivalled Mussolini's, he was banished to Libya to be its Governor. On June 28, 1940, Governor Marshal Italo Balbo was shot down and killed by his own anti-aircraft guns over Tobruk, just after an RAF raid had begun, and Balbo disappeared into the mist of history.

MAIL AND PASSENGER SERVICE

FROM THE BEGINNING OF man's flight—whether instinctively or consciously—a general purpose appeared to have been high in inventors' minds: war. Not much progress had been made in using the flying machines for mass transport of cargo and passengers. As the activity at Harbour Grace had suggested, the attempts at perilous Atlantic crossings by air had not only been daredevil wishes to set records, but also to bring the Big Pond under Man's control. It took a horrendous world war and total effort to make Atlantic crossings an every-day reality.

Harbour Grace's flirtation with airplanes and Atlantic crossings took a twist in 1922, when a restless Australian inventor—who played important

rushing, hull more visible, shining wet, emerging, then trailing comet-bursts of flashing liquid silver thrown by the slipstream they were airborne. What a grand thing! The first ones climbed in circles until all were up, up into the fresh air, and then they joined, set a heading directly east, and gradually the magnificent men in their armada droned away until… silence. The watchers looked about, emerging from a spell. What a day!

The visit was one of the most important events in the town's history, so much so that the coastal road fronting Shoal Harbour was renamed Balbo Drive. Chicago had named a street after Balbo, as well. In New York, there was a ticker-tape parade and President Roosevelt entertained Balbo for lunch. Balboa was a hero in

AIR MAIL

roles in two world wars—landed in the town's harbour from St. John's, on Feb. 24, with the first airmail delivery in the Crown colony. Born in 1894, in Goorganga, Queensland, Australia, Sidney Cotton went to England at the age of 21 to join the Royal Naval Air Service. After only five hours solo flying, he was sent to the Western Front and managed to survive both squadron training and the air war, emerging a major. He was said to have invented rear-facing guns on his Sopwith 1½ Strutter fighter, to defend his back (his "six" in later jet-jocky parlance).

After the war ended, Cotton moved to Newfoundland and earned a living doing aerial surveys, photography and locating game and seals for hunters. In 1920, he established a seaplane base at Botwood, on the mid-northern coast. His Botwood base eventually caught the eye of British planners looking for potential sites for a regular North Atlantic mail and passenger service using multi-engine flying boats. This followed a 1935 agreement between Canada, the United Kingdom, the Irish Free State and Newfoundland (and also inspired construction of Gander airport whose later massive maze of runways became a major trans-Atlantic air terminal).

Imperial Airway's *Caledonia* had stopped at Botwood on the way. Soon Imperial Airways and Pan-American Airlines flying boats confirmed the 1935 international agreement and, by 1939, scheduled mail and passenger flights, including the romantic Boeing "Yankee Clippers," regularly were spanning the Atlantic, and the Sikorsky S-42 Clippers were serving the South American routes. These airlines, and others, made Botwood a regular stopover. In July, 1939, the Pan American flying boat *Clipper III* took off from Botwood to initiate regular airmail service between Canada and Britain, and carrying passengers. Passing it from the other direction was Imperial Airways four-engine Caledonia headed for Botwood, the 3,204 km one-way trip averaging 15 hours and 9 minutes.

Long-flight ocean passenger service in the beautiful Boeing 314 "China Clipper" and "Dixie Clipper" flying boats of Pan American Airways were providing 2,000-mile service for both trans-Pacific and trans-Atlantic schedules by 1939. They filled a void left by the Zeppelin airships, which stopped service in 1937 with the loss of the *Hindenburg*. PAA flew its last Clipper service in 1946, after operating through the war on military duties.

The Second World War effectively ended the clipper services. Now, a huge wartime advance for TCA in Atlantic passenger travel beckoned. In 1943, the Canadian Government Trans-Atlantic Air Service had been set up, and TCA pilots began flying Avro Lancaster bombers built in Canada by the new Victory Aircraft Company using two routes, one through Gander and the other via Goose Bay. These Lancaster XPPs were configured as transports, rather than bombers, and flew flights averaging just over 13 hours. They were to provide regular mail service to overseas Canadian servicemen, and transport key personnel on wartime assignments.

In 1944, 3,726 aircraft, such as Canadian-made Avro Lancasters, de Havilland Mosquitos, Boeing Flying Fortresses, Lockeed Hudsons, Liberators were flown across. Allied fighter, bomber, maritime patrol and transportation squadrons received 9,027 aircraft via ferry flights throughout the war. But the brilliant end result was that the Lancasters ushered in the era of regular passenger air travel across the North Atlantic and, eventually, officially became part of Trans-Canada Air Lines (Atlantic), in 1947. TCA began scheduled service in 1947 between Montreal and London, flying Canadair DC-4 North Star airliners.

AS WAR LOOMS, THE MOSCOW TO MISCOU FLIGHT

DURING THE TRANSITION to jet training at Chatham's operational training unit we criss-crossed a good deal of New Brunswick terrain at high altitude, and the day I stretched northeast to Miscou Island jutting into the Gulf of St. Lawrence, I was ignorant of the drama that had played out down there 14 years earlier—in 1939. Joseph Stalin's favourite air force general had crash-landed a bomber on the beach at Miscou. And, within a few months, I would be patrolling European skies ready to tackle up-to-date jet bombers that same general had helped to develop—the core reason NATO fighters were stationed on the Continent.

In 1939, fascist forces were playing havoc in the world. They were

 Moskva was a small red Illyushin TsKB bomber. The Russians planned a brilliant attention-grabber—a world-record-smashing trans-polar, non-stop flight from Moscow to the 1939 New York World's Fair—but ended up on a New Brunswick beach. RCMP Constable Theriault and unidentified person on the bomber. *Photo: Provincial Archives New Brunswick/ Nicholas Denys Historical Society*

singing war songs in England: "Hang Out the Washing on the Siegfried Line," "The Last Time I Saw Paris", and "Lili Marlene"; in Germany, they were singing, "Wir fahren gegen England", "Bomben auf England" and "Lili Marlene". Joseph Goebbels, Hitler's propaganda minister, presented his enemies with a treat: Charlie and His Orchestra, aping American swing music but substituting Nazified lyrics in jazz hits such as "Thanks For The Memory," "The Shiek of Araby," "Indian Love Call" and, of course, "Lili Marlene," broadcast on medium and short-wave to Canada, U.S. and U.K.

Canada had sent out its first sea convoy bearing materiel for a Britain at war; would declare war on Germany; formed Defence Industries Ltd. to make explosives and munitions; the federal government would assume wartime powers after proclaiming the Wartime Measures Act. The Ottawa Car Company had begun building heavy bombers.

North Americans still were shaking off the dust of a devastating Depression in 1939 when the United States decided to build a democratic utopia's example of a Town of Tomorrow and called it the New York World's Fair. To this fairy-tale display it attracted the best of a totalitarian country, the USSR, which went all-out to propagandize. The Russian community of states erected an impressive pavilion, and maybe equalled the Americans in their pizzaz. Massively dignified, the exhibition featured a giant statue of a Russian worker, a flowing Red Star held aloft proudly. Displays were of socialism's industry and agriculture, work and recreation.

But the Russians also knew about showmanship that sparkles, and planned a brilliant attention-grabber—a world-record-smashing trans-polar, non-stop flight from Moscow to New York, to coincide with the Fair's opening ceremonies. What better way to mark the $100 million event? The plane, one of their top new low-wing bombers, would be piloted by a hero of the Soviet Union, holder of more than a dozen world aviation records. The plane would fly more than 4,000 miles, passing over Norway, Iceland, South Greenland, New Brunswick, the coast of Maine, Boston and, finally, swoop down to launch the World's Fair extravaganza with a spectacular landing at Floyd Bennett Field, in New York.

Everything went perfectly, until… the pilot had compass trouble and had to land on a rough, flat area near the beach of Miscou Island, off the coast of northern New Brunswick. So the try for a world record became the "Moscow to Miscou flight," wrote Monsignor Donat Robichaud, whose story about the 1939 flight appeared in *The Atlantic Canada Aviation Museum Newsletter*. And once again, global attention fell on Atlantic Canada where so many flights had ended during the thrilling 1930s, the decade of world record attempts.

Brigadier-General Vladimir Konstantinovich Kokkinaki, 35, tall, handsome, a professional-standard boxer and weight-lifter, was famous as a high-altitude flyer, reaching 11,458 metres in 1935. In 1939, in preparation for the New York flight, he climbed to 14,660 metres (48,097 feet) in an open cockpit. He

Major Mikhael Gordienko, radio operator and navigator, left, and Brig-Gen. Vladimir Konstantinovich Kokkinaki, pilot. Compass trouble turned a Soviet propaganda triumph into a pratfall. This is a copy of the 50th anniversary Russian postcard to mark the flight in 1989.

figured the Moscow-New York flight on the "great circle route" would take 25 hours. Kokkinaki dumped the aircraft's life raft, stating: "We intend to fly to America, not paddle there!" The radio operator and navigator on the mission would be Major Mikhael Gordienko. The plane, Moskva, was a small, red bomber, a twin-engine (each 1,000 hp) Illyushin TsKB, with additional fuel tanks, a type used by Russians fighting for Loyalist Spain. The ground speed for the 4,000-mile trip was estimated at 165 to 225 mph.

Weighing in at 12.5 tons, *Moskva* took off from Moscow at 9 a.m. New Brunswick time into unexpectedly tough headwinds, but found it was on time over Iceland. By mid-afternoon, storm clouds forced the pilot to climb on instruments to 30,000 feet. The crew's clothing protected them from the extreme cold, but the oxygen supply was rapidly being used up. When they left cloud, they were above the Gulf of St. Lawrence, and night was approaching. The Russian ambassador in Ottawa had asked CBC radio to help guide the bomber's crew through Canadian air space by broadcasting a short message in Russian. Gordienko picked up the message, but his radio compass had stopped functioning in the extreme cold. The record-seekers couldn't figure out where they were.

Kokkinaki knew he had to land before nightfall. He descended and, at 8:55, landed on an unknown island after about 22 hours of flight. Gordienko immediately contacted Moscow. They could have made their destination; there still were 330 gallons of fuel on board, enough to fly another 940 miles, and New York was just 650 miles farther on. Authorities in Moscow radioed the control tower at New York's Floyd Bennett Airfield that Moskva had made an emergency landing, "south of the Hudson Bay."

Kokkinaki almost made a perfect forced landing. With wheels partially lowered, he selected a strip of terrain between sand dunes along Chaleur Bay, and trees inland. He was leaving a straight line of tracks on the wet ground until one wing snagged a small tree that swung them around. But the flyers had become the first to complete a northern flight from Russia to North America.

The few people living in the landing area had heard the engines of a low-flying plane near their church. The village priest, Father Ernest Chiasson, recalled: "I went outside to see what was happening. The plane circled the church ten times as if trying to find a large field in which to land. Then it flew a bit farther, maybe to avoid the house, and find another site. It was twilight. Farther away, we saw it fly over the plain with a landing light to inspect the terrain. Finally, the motor died and we heard the sound of a crash, then silence."

At nearby Shippagan, the telephone operator, Antonine Robichaud, had read about the proposed Russian flight but hadn't paid much attention. At 8:50 p.m. she got a call asking about a plane crash on the island. She called around and was told by several residents that they had heard a plane, and then a crash. "My mother," she was quoted in the museum newsletter, "sent a telegram to radio station CFCY in Charlottetown, P.E.I.: 'Plane sighted 8:50 p.m. thought to have landed. If wanted will wire further particulars soon as available by wire or phone. Advise.'" Night fell. Many Miscou residents had seen or heard the plane, and had an idea of the direction of the crash. Brothers Lawrence and Bert Vibert walked and searched in the general direction for two hours. Aided by flares set off by the airmen, they located the aircraft. But, "as they approached the plane, they were greeted by two Russian aviators, revolvers in hand, who refused to let them anywhere near the aircraft. It was impossible to communicate as the pilots knew neither French nor English." Lawrence indicated as best he could that he would return with food and help. The round trip took more than three hours.

They reported that one of the airmen appeared to be injured. A doctor in Tracadie was called. The telephone operator had Lawrence Vibert write down the names of the aviators and return to the crash scene. "He showed the pilot standing guard the paper, which said KOKKINAKI, and he pointed to his injured comrade, lying on a makeshift bed near the plane. Lawrence had also brought food. By signs, he realized that the aviators wanted the military advised of the situation." Reassured, the Russians used a map to show they had been headed for New York. They were warmly dressed and wanted to sleep by the plane because a gas tank had been ruptured and they feared fire.

Once CFCY got Madame Robichaud's message, the word spread and ambassadors, international media, world's fair authorities and emergency services were all trying to get more information. The only contact was Antonine Robichaud in her Shippagan telephone office. "They were calling from everywhere. From London, Boston, New York, Toronto, Moscow: 'Please get this wire through…Get General Kokkinaki on the phone…Can I speak to Major Gordienko…This is the National Broadcasting Company…Hurry!'" A 24-hour line was maintained with Moscow along a system of ground and marine cables passing through Poland, Germany and England; wireless from England to Yamachiche, Quebec, by cable from Yamachiche to Montreal, to Shippagan and, finally, underwater cable to Miscou, via la Pointe Brulee.

An RCMP constable and a doctor rented a small rowboat early Saturday morning to cross from Shippagan to Miscou. They walked across the island and reached the plane that afternoon, to find that a doctor from New York already had arrived. Toronto and Montreal newspapers asked Cyril Mersereau, of Bathurst's *The Northern Light*, to take a photographer to Miscou and get a story ASAP. After a long drive, then a trip by small boat, and walking over ice floes to Ile Lameque and Ile Miscou, they met Father Ernest Chaisson, who drove them to the crash vicinity and they walked in.

Now news reporters, mostly from the eastern U.S., were flying into Moncton, the nearest commercial airport. At sunrise Saturday morning, a Canadian Airways Fox Moth managed to set down on the island, not far from the crash, carrying a *Canadian Press* reporter, photographer and an aircraft mechanic. A Waco biplane, rented by the *New York News* appeared over Miscou and a *New York Times* photographer took shots from the air. Harold S. Vanderbilt, of America's Cup yachting fame, flew into Moncton in his 14-seat Lockheed 14, carrying members of the Russian embassy who had brought passports for the stranded aviators. Then a Grumman Goose amphibian rented by the Russian embassy landed at Moncton, later a Beech 17 staggerwing, from New York, and a Steerman, from Boston, carrying *Associated Press* and *International News Service* reporters and photographers.

Now the trick was to fly the Russians from the crash scene. The Grumman landed in Miscou Harbour, six miles from the crash, but ice floes and approaching dark forced it back to Moncton. An amphibian from Shearwater's RCAF station, with an interpreter, landed at sea three miles away, and its passengers got ashore in a small boat, then walked through snow and mud to the site. Finally, the Fox Moth got in to the site, took the Russian navigator Mosienko to Moncton, then went back and got Kokkinaki. •

>=

A pride of Bristol Bolingbrokes. Built in Canada, they were a prime multi-engine training aircraft for the British Commonwealth Air Training Plan in the 1940s, long remembered by heavy-bomber crews. Many remained strewn in pieces, or gracing farmers' fields and farmyards.

CHAPTER FIVE

THE SECOND WORLD WAR

WE COLD WAR FIGHTER PILOTS swarmed in our Sabres from a well-used former RAF base, criss-crossing the Midlands and the fens of East Anglia looking for action, knowing that we were pretty well the only show in town. Swept-wing Canadair Sabre jet fighters of 410, 439 and 441 (F) Squadrons were like a shot of adrenalin at North Luffenham, designated No. l (Fighter) Wing, RCAF, "Spearhead of NATO." The wing's three squadrons, and nine additional squadrons soon to be located on the continent, would claim to be the most aggressive pilots in the air throughout the British Isles and continental Europe.

Just six years after the war, we had moved onto a station that had played a stalwart role in that war. From these dark runways had climbed into the night skies Vickers Wellingtons, Handley Page Hampdens and Halifaxes, Avro Manchesters and Lancasters, Armstrong-Whitworth Whitleys, Short Sterlings, and Douglas Dakotas, some towing Horsa and Hamilcar gliders. The station had lost sixty bombers on operations.

In 1951, we Canadians had set up shop on Luffenham Heath that nestled just south and east of Rutland's centre, near Eyebrook Reservoir—a handy navigation reference used gratefully (if there was moonlight) during the war by night fighters also based there—Bristol Beaufighters and Canadian-made wooden wonders, de Havilland Mosquitoes.

PREPARING FOR WAR

AS THE STRUTTING AND PUSHING got nasty in 1939 Europe, Canada was caught with its pants down. The Great Depression had brought any aircraft manufacturing hopes to a standstill and the air force was in disarray. There were only eight permanent squadrons to take war stations in Canada that year. "None of the units was fully manned or equipped," says W.A.B. Douglas in

The Creation of a National Air Force. "As of 5 Sept., the air force had only 4,153 officers and airmen, far fewer than its authorized establishment… Of the fifty-three aircraft able to take their place on active service, including eight on the west coast and thirty-six in the east, many were civil types converted with floats for patrol work and most of the others were obsolescent."

Ottawa loosened the purse strings a little in 1937-8, which allowed the purchase of several more aircraft from Britain, though all were obsolete, or near to it. One bright spot: with future needs in mind, Britain decided some of these would be built in Canada, allowing a measure of hope after the lean years for a developing aircraft manufacturing sector. Bolingbroke bombers were built by Fairchild Aircraft of Canada Ltd., and Lysanders were manufactured by National Steel Car Company. De Havilland's first manufacturing job was to fill a Canadian order for 26 Tiger Moths for the RCAF. These would be delivered by the time war was declared. Fighter aircraft were a low priority for Canada's scheme of war (homeland protection), but a few Hawker Hurricanes were bought and entered service in the spring of 1939.

Then Britain provided added stimulus to the Canadian aircraft industry. It arranged a central contracting company that would assemble British-designed aircraft for the Air Ministry. Six firms combined to form Canadian Associated Aircraft Ltd., which assembled airframes from parts and components supplied by the six. Early on, eighty Hampden bombers were ordered, with a promise for another 100. And 40 Hurricanes were ordered from the Canadian Car and Foundry Company. These orders led to much larger ones later. Manufacturing of cars in subsidiary plants (Canadian Car and Foundry; National Steel Car; Ottawa Car Manufacturing Co.) had been a big Canadian industry since the 1920s, but production had ground almost to a halt during the Depression. The manufacturing plants would prove indispensable when converted to war truck (soon to be a huge Canadian enterprise) and aircraft production. At this time, Douglas McCurdy left Curtiss-Reid and joined the

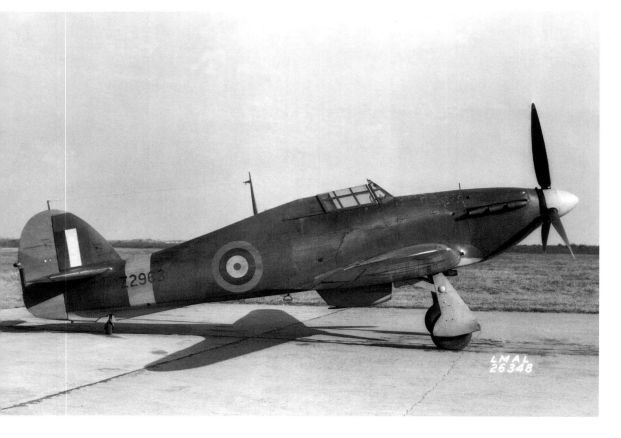

Fighter aircraft were a low priority for Canada at first, but a few Hurricanes had been bought for homeland protection by 1939. As war progressed, a huge demand saw Canadian factories turning out hundreds of the hardy and effective fighters.

Canadian war effort as supervisor of purchasing and assistant deputy director-general of aircraft production, eventually earning an Order of the British Empire. He remained in that position until 1957, when he was installed as lieutenant-governor of Nova Scotia.

One of the greatest battles fought by Canadian air planners during the war was against the U.S.-Britain coalition. Canada might be on the same side, but when it came to designing its own war policies, and carrying them out, they had to bow to the wishes of the Big Boys. The problem was that Canada didn't have many aircraft plants, and no original designs, to create a national air force. Even as its plane-building ability increased, it didn't manufacture aircraft engines with which to power the planes. Canada developed policies for home defence, particularly on the coasts (where important anti-submarine duties were shaping up), but grandiose hopes to assemble many squadrons, with suitable aircraft, were always dashed by the U.S. and Britain, who had their own priorities, and held all the strings in aircraft and engine production.

"Canadian airmen were caught between their government's insistence that Canada be well defended, their own aspirations to construct a respectable national air force, and the fact that they did not control the resources to meet either objective," wrote W.A.B. Douglas in *The Creation of a National Air Force*. Air Marshal L. S. Breadnen's staff was hard-pressed to raise 20 undermanned squadrons while he wrote about his dream of a 49- to 65-squadron air force. Air staff policy sometimes made little sense, and the Home War Establishment had to improvise.

Nonetheless, first steps were taken. No. 10 Squadron was formed in September, 1939, for submarine surveillance. It flew Wapiti, Digby and Liberator aircraft, recorded 3,414 sorties, set a record of 22 attacks on U-boats, sinking three. Unofficially, but proudly, it was known as the North Atlantic Squadron. Eventually, with eight maritime patrol squadrons and 78 aircraft on the Atlantic seaboard, the RCAF carried out extensive air surveillance of the Northwest Atlantic. Canadian squadrons with the RAF, patrolled from the other side of the water. Canadian aircraft were involved in the sinking of 51 of a total 785 submarines sent to the bottom during the war.

In early 1941, under an agreement between the U.S. and Canada, Canadian Vickers Ltd., at Cartierville, Quebec, and Boeing Aircraft of Canada, at Vancouver, began production of the PBY-5 and PBY-5A. In early 1942, Canadian Vickers produced them. A total of 254 Catalina and Canso (amphibian) flying boats were used by RCAF units on the west and east coasts. During the Battle of the Atlantic, PBYs on anti-submarine patrol were often successful against German U-Boats. F/L David Hornell received the Victoria Cross for his actions after engaging a German U-Boat in the North Sea with one of these aircraft.

Heavily-armed Spitfires, flown by pilots often just out of their teens, teamed up with Hurricanes to defend the homeland.

BCATP SET THE TRAINING STANDARD

IN-SERVICE TRAINING OF pilots and other aircrew was developed to a fine art during the Second World War when Canada became, in the words of the U.S. president of the time, Franklin Delano Roosevelt, "the aerodrome of democracy."

Canada's former "in-house" expertise in aircrew training, as has been pointed out, was born during the First World War, when Britain counted on a recruitment and training organization here to provide pilots for the Royal Flying Corps. Then, when war clouds rolled in again in 1939, Canada agreed to take on the British Commonwealth Air Training Plan (BCATP), and pay for most of it. Prime Minister Mackenzie King initially hoped this would provide the bulk of Canada's contribution to the war effort. This massive undertaking resulted, within a few months, in the levelling of thousands of acres

FLYING BLIND

HEAVY BOMBERS AND NIGHT FIGHTERS depended on flight instruments, and navigation aids were developing monthly. In the early days of aviation, "blind flying" had been a term referring to the loss of essential eye recognition of "up" or "down" after meeting unforgiving conditions such as thick cloud, snow, night with no moon or star sightings. The horizon is the classic reference, and the ability to see that natural feature is what kept pilots from spinning in. No one knows how many fliers have lost control and spun in—many were lost while trying to fly across the Atlantic in hazardous conditions—because they weren't trained to read their flight instruments or, as in those experimental days of flying, there weren't many sophisticated gadgets on the instrument panel. To them, blind flying meant, well, taking chances in bad flying conditions. It could be deadly when "seat-of-the-pants" didn't work.

"Blind flying" was a magical phrase, a conjuration emotional and exciting to boys of the time. It meant daring and adventure, because pilots were the ones who did it, and knew unfathomable things that always got them out of a pickle and safely back to base. There was no television, and kids listened to radio serials about flyers, and read about them in newspaper comic strips (Smilin' Jack), and watched them at the Saturday movie matinees. A suitable title for a movie about blind flying would be "Lost Horizon." The plot would include the wisdom of Van Sickle's Modern Airmanship: a pilot "must learn to rely on all the instruments, to the degree that any bit of information he needs can come from the maximum number of instruments, and failure of a given instrument does not leave him unable to continue safe flight."

Through the first half of the 1940s, hundreds of harried pilots—flying out of North Luffenham and scores of other airfields in England—learned the hard way what "failure of a given instrument" and blind flying could be about. Their wreckage, and dead or damaged bodies were strewn about without discretion.

These victims predominantly were from Bomber Command, that Allied aerial force described so vividly and honestly by one of their number, D. Murray Peden, Q.C., D.F.C., in his still popular book, A Thousand Shall Fall: "Over 55,000 aircrew were killed serving in RAF Bomber Command, thousands of that number being fellow Canadians. The crews faced formidable odds, odds seldom appreciated outside the Command. At times in the great offensives of 1943 and 1944 the short-term statistics foretold that less than twenty-five out of each one hundred bomber crews graduating from Operational Training Unit would survive their first tour of thirty operations. On a single night Bomber Command lost more aircrew than Fighter Command lost during the Battle of Britain. Yet the crews buckled on their chutes and set out with unshakeable resolution night after night."

The Lancaster at Hamilton's Canadian Warplane Heritage Museum—one of only two in the world flying—is dedicated to the memory of Pilot Officer Andrew Mynarski, and bears the colours and markings of his aircraft that flew with 419 (Moose) Squadron. *Photo: Courtesy Airforce Magazine*

of farmers' fields for transformation into operational schools. At a cost of more than $1.6 billion (an astronomical sum then), 94 schools operated at 231 sites across the country. More than 10,000 aircraft and 104,000 men and women were involved. In all, 131,550 pilots, air gunners, navigators, bomb aimers, wireless operators and flight engineers from 40 nations were graduated.

Testament from an RCAF officer: "Because of its nature, the British Commonwealth Air Training Plan is unique in any military form. There never had been, in any one country, quite that concentrated a gathering of people to train, and be trained, and to create the whole machine. It was an entirely new thing, anywhere for anything, military or commercial. Canada had the broad horizon of air and land, but little else. It had never done anything of this scope, to have to build the plant, and train the men to fly, who would then train thousands of other men to fly."

The wisdom of setting up and supporting flying clubs in Canada to train civilians in the 1920s paid off

A gaggle of Mark 4 Harvards above grain fields, a familiar scene for a quarter century over the Canadian Prairies. *Photo: Courtesy Airforce Magazine*

De Havilland Tiger Moth in which thousands of early-war Canadians learned to fly for the RCAF. *Photo: Courtesy Airforce Magazine*

when war clouds gathered in 1939. Britain declared war on Germany on Sept. 3, 1939, and within days the country's 22 flying clubs were organizing its members to train pilots for the RCAF. "Fortunately, in the preceding years the RCAF had encouraged the clubs to adopt RCAF training systems and standards (James N. Williams, The Plan)…In order to operate Elementary Flying Training schools for the RCAF, each of the clubs was required to re-organize into a commercial company, and each of these companies was then required to post a bond to indicate good faith and stability."

John Gillespie Magee learned to fly with the Canadian air force during the war and then wrote the 14 lines of "High Flight" that were adopted by the RCAF as its official poem. Jonathan F. Vance wrote a book on the history of flight in this country in 2002, and by titling it High Flight, set himself an impossible goal in Canadianship—and met it. In this even-handed and evocative work, Vance wrote, "When the RCAF ballooned from a few thousand members in 1939 to a quarter of a million six years later, it fundamentally changed the relationship between aviation and the general population. During the First World War, only a select few Canadian families could boast an airman in their ranks; by 1945, hundreds of thousands of families had members in the air force or air cadets."

THE HALIFAX

IT WAS ANOTHER CLOAK-AND-DAGGER job facing the crew of X for X-Ray NA 337 when the four-engine Halifax bomber lined up on the runway at Royal Air Force Station Tarrant Rushton, near Dorset's English Channel coast. The evening dusk of April 23, 1945, was suddenly punctured by a green Aldis Lamp flash and the pilot shoved the throttles forward for a thundering take-off at 7:51 p.m., three minutes behind schedule. The rear gunner, sitting in the coldest part of the aircraft, settled in with his thermos of coffee filled to the top for the trip to Norway — always a nine- to 10-hour grind. They were due to touch down at 6:14 a.m., next day, but the Halifax, and the thermos, weren't seen again until 50 years later — the thermos still containing the gunner's coffee.

NA 337 was a modified Halifax Mark AVII, of 644 Squadron. The squadron's aircraft did "special duties," mainly for the Secret Operations Executive (SOE). There was no upper turret (hence just a six-man crew) and the dome usually found on the underside had been removed and the space fitted with opening doors for dropping "packages" and "Joes" (agents) to Resistance Forces behind enemy lines.

This was NA 337's fifth mission since being built by Rootes Securities at a shadow factory abutting Speke airfield, outside Liverpool. It had been delivered to the RAF early in March, 1945. Its first mission involved towing a Hamilcar glider containing a Dodge truck and a 17-pounder gun on Operation Varsity, the massive airborne assault across the Rhine River into Germany. The next four missions all were for the SOE, one over Denmark, the others over Norway.

The crew on this flight — Flying Officer A. Turnbull, DFC, 27, pilot; G.R.Tuckett, 23, air bomber; Flt. Sgt G.A. Bassett, 34, flight engineer; Flt. Sgt. A. Naylor, 22, wireless operator/air gunner; Flt. Lt. W.R. Mitchell, 23, navigator; Flt. Sgt. T. Weightman, air gunner — was on its eleventh mission. Their first mission together had been Oct. 10, 1944, when their Halifax was damaged by enemy fire. Of their missions so far, two were bombing runs over Germany; eight were for SOE.

The twin-tail Halifax cruised at 225 mph through the dark towards Norway, powered by four Bristol Hercules engines, each producing 1,165 hp. On board were two SOE "packages" and 13 SOE containers for Milorg, the Norwegian resistance movement. The dropping mission was code-named Crop 17. The drop target was in the mountains near Gruce, 80 miles northeast of Oslo. After picking up good signals from the reception committee on the ground, the crew successfully made the drop some time after midnight. The aircraft then turned for home, flying almost due west. Over the south end of Lake Mjosa, enemy air defences spotted NA 337 flying at 700 feet and passing over a bridge at Minnesund, on the key Trondheim-to-Oslo railway line, at 1:45 a.m. German anti-aircraft guns opened up and 20mm shells ripped into the Halifax, setting a fuel tank afire near the outer starboard engine. The propeller was feathered, to cut drag. Skipper Turnbull banked the bomber onto a northerly heading to fly just west of Tangen, Stange, and a little beyond Hamar, before reversing direction and heading south again to ditch on the partially-frozen Lake Mjosa.

The impact tore away the other three propellers, one vertical stabilizer, the port wing tip and the airplane's nose. The tail section and 17 feet of fuselage snapped off. All but one of the crew, who had gathered at the

Halifax NA 337, similar to the one below, was shot down and sank in a Norwegian lake, in 1945. Fifty years later, the aircraft was brought to the surface and an air gunner's thermos flask was found, still full of coffee.

ditching position in the centre of the aircraft, got out. The survivors managed to inflate their Mae West life jackets, but the freezing water took its toll. When two Norwegian teenagers came out in their boat to look for survivors at first light, only rear gunner Flt. Sgt. Weightman was found alive, lying on top of an overturned dingy. The flight engineer, Flt. Sgt. Bassett, was never found. The dead were later buried in the Commonwealth War Graves Commission plot in Nordre Cemetery, Lillehammer, Norway.

Halifax NA 337 had disappeared through the ice into the black depths of Lake Mjosa.

Nearly half a century later, in 1994, a plan to raise NA 337 was conceived by the Halifax Aircraft Association (HAA), led by Karl Kjarsgaard, of Ottawa. Kjarsgaard was a 767 pilot for an international airline, which allowed him to search out recovery projects while travelling around the world. The HAA mainly comprised eight veteran wartime airmen who had flown or worked on the Handley Page Halifax bomber. They each had their own special reasons for working on the airplane restoration.

"It's a labour of love," said 77-year-old Morris Ducarme, an ex-air force metal worker, in an interview before the restoration was completed. "We worked together in uniform and we work together here." Another senior volunteer on the project was Lloyd Wright, 81, a former Halifax pilot. Wright's wife joked, "We used to winter in Florida until that other woman (NA 337) entered our lives."

First, though, the members of the HAA were determined their project would tell the world about Canada's heritage in the air. Their goal was to restore the Halifax bomber for display in the Royal Canadian Air Force Memorial Museum at Trenton, Ontario. Lt. Col. (Ret.) Joe Bourgeois, chairman of the museum's board, said the Halifax would be "a proud memorial to the 10,000 men and women in the RCAF who have given their lives for their country." One of the founding members of the HAA was John Stene, a Norwegian airman who fled his country when the Germans invaded in 1940. Ultimately, he ended up in Canada. When the HAA was formed, it was Stene who negotiated with the Norwegian government for permission to salvage NA 337.

Another member of the association, Tore Marsoe, was sixteen on the night he heard the crippled Halifax fly over his hometown of Hamar, Norway. In the early 1980s, Marsoe and a friend had been conducting a futile search of Lake Mjosa for the lost Halifax. Then new information encouraged them to check another area of the lake with sonar, where they soon found the wreck. The main fuselage of NA 337 was located at a depth of about 750 feet (240 metres). The tail section was some distance away, in 660 feet (200 metres) of water.

In June 1995 a Norwegian salvage company was contracted to raise the wreck, and on Thursday, Sept. 7, 1995, Jeff Jeffery, DFC, the HAA's then president, saw NA 337 for the first time. He stared at the outline of the once 34,000-pound airplane, with its 104-foot wingspan, raised to just beneath the surface. Both shattered ends of the fuselage had been secured to prevent loss of artifacts. The beaching operation was dramatic. A winch slowly towed the aircraft's remains towards shore. "First the wingtips appeared," Jeffery said; "then the balance weights for the ailerons; then the engine cowlings and the bent prop on the outer starboard engine. Finally, the fuselage and the upper surfaces of the wings appeared.

"When we had stopped to hook the beach winch to the sled, I went out on a work raft. I stood on the wing and looked into the cockpit (and put my arms around it and cried). What a wonderful sight! The throttles, pitch controls, gauges and all were a little muddy and wet, but all there."

Once on shore a 12-man crew from the Canadian Air Force's Number 8 Aircraft Maintenance Squadron, based in Trenton, Ontario, flew in ready to clean, dismantle and crate the aircraft remains for shipment to Canada. Headed by Capt. Doug Rutley, the CAF crew moved the airplane's remains from the rocky beach and prepared them for shipping. "This was the most challenging and rewarding assignment of my career," said Rutley. "Everyone on the team gained something from this experience. There were certain moments for each of us, which made us all proud to be Canadians and to be able to play such an important part in preserving our military heritage."

The job took two weeks to complete and by the middle of December that year the Halifax was ready to be flown out. When all was done, it took eight Canadian air force CC130

After years of work, a restored Halifax aircraft, NA337, was moved in 2004 to an extension of the RCAF Memorial Museum at Trenton, Ontario. The airplane, like the one above, is the only complete Halifax in the world. *Photo: Courtesy Airforce Magazine*

Hercules loads to transport the remains to Trenton. Although NA 337 was an RAF aircraft, the HAA wanted to restore the Halifax bomber as a distinct part of Canadian history. After all, "Hallies" made up 15 Canadian squadrons during the Second World War. But unfortunately, at the end of the war the British had all wartime Halifax bombers cut up for scrap, not one being saved intact as a museum piece.

After many more years of dedicated labour by dozens of skilled retiree volunteers, working under the supervision of volunteer Lt. Col. (Ret.) Bill Tytula, the restoration of NA337 at the RCAF Memorial Museum was completed. The fuselage, tail section, inboard wing sections and inboard engines were assembled and raised to flight position. All four engines were installed. The forward fuselage and cockpit were attached, a mid-upper gun turret acquired, and replica machine guns manufactured. And, interestingly, a former Halifax navigator personally paid for the upholstering and refinishing of crew stations.

A number of enthusiasts around the world also pitched in to assist the restoration project. A Canadian living in California saw a Mk.IV bomb sight at a San Francisco yard sale, bought it and delivered it personally. Another visitor arrived from the U.K. with a complete antenna-switching unit, apparently having noticed during an earlier visit that the original was in bad condition. In addition, several years of work and scrounging went into finding enough components to create three new propellers. Parts have come from as far afield as Holland, Scotland, Ireland and Edmonton.

On October 22, 2003, the RCAF Memorial Museum broke ground on a $4.8-million, 48,000-square-foot expansion, which included space to house the restored bomber. On October 16, 2004, NA 337 was moved into the new display hangar with appropriate ceremony. The grand opening of the entire museum addition was held a couple of years later. The last near-perfect example of a vintage Handley Page Halifax bomber now stands high and proud in its new home. It is a fitting testimony to the bravery of Canada's young airmen, as well as to the vision and dedication of the once-young Canadian men who

REMEMBERING OUR FLYERS

WE WERE WAITING IN THE HAMILTON terminal for a connecting flight during a trip to St. John's a couple of years ago, and opted to make good use of the time by visiting the Canadian Warplane Heritage Museum. What a great decision! The history of Canada's air force is all there, intact, cared for. Respected! And no other Canadian aviation museum has so many of its exhibits airworthy and frequently flown. Besides the Lancaster, two other aircraft—the 1930s Boeing Stearman open-cockpit biplane trainer and the North American (Canadian Car and Foundry) Harvard trainer—also take up passengers. Two flights are offered: a bird's-eye view of Hamilton and the Lake Ontario shoreline; or Niagara Peninsula and the Grand River area.

Hamilton's Lancaster is dedicated to the memory of Pilot Officer Andrew Mynarski, and bears the colours and markings of his aircraft that flew with 419 (Moose) Squadron. On June 13, 1944, a night fighter shot his bomber down in flames. While it was going down, Mynarski, ignoring his own safety, seared by fire, tried to free the trapped rear gunner. Unable to help him, he saluted, before jumping. The gunner, in one of the flukes of war that happen, survived; Mynarski, his parachute aflame, fell to his death. He was awarded a posthumous VC.

For aviation enthusiasts, the museum at Hamilton International Airport is a treat, a 108,000-square-foot delta-winged building opened in 1996. Chock full with almost every aircraft flown by Canada's air force pilots—including a diminutive First World War Sopwith Pup biplane—the museum is open daily, closed only on Christmas, Boxing and New Year's days. It is affiliated with the Canada Aviation Museum, Ottawa. There's an appealing atmosphere of respect and remembrance of a past through aviation displays, memorabilia and photos of Canada's leading aviators. Muted background sound is the nostalgic music of the wartime years, ballads and Big Bands.

On the upper level, we watched volunteers at work building a replica of the *Silver Dart*, that first plane to fly in Canada in 1909, product of the organization run by Alexander Graham Bell. All operations (mostly by volunteers) are under one roof: aircraft rides (all pilots have commercial licences); gift shop, cafeteria, guided tours, aviation library; vintage aviation movies in the Rolls Royce theatre. Meeting rooms can be rented. Private dinners, reunions or theme events can be arranged, guests dining and dancing in close conjunction with propellers or jet intakes.

The museum is just a five-minute stroll from the airport passenger terminal, easily identifiable by the 1,400 m.p.h. F-104 Lockheed (Canadair) Starfighter at the entrance, standing on its tail, blasting skyward. Visitors can test their flying skills in flight simulators. They can sit in the cockpits of trainers and fighters. They can get close up to all the colourfully-painted and camouflaged aircraft. Using remote controls, they can manoeuvre a model Hawker Hurricane fighter (remember The Few in the Battle of Britain?) or raise and lower flaps and undercarriage of a real Lockheed (Canadair) T-33 jet trainer.

Why pay $1,400 to fly in an old bomber? Well, Hamilton's the only place in the world where you can. The Royal Air Force has the only other flying Lancaster, but it doesn't offer rides to the public. There's a mystique involved, particularly for older people with memories of the aircraft that played a huge role in winning the Second World War. Thousands of Canadian boys from coast to coast were Lancaster crew members—gunners, navigators, bomb aimers, pilots—and thousands of them died during raids over occupied Europe. People recall the exploits of Guy Gibson and his Dam Busters, and the movie that followed, featuring the Lancaster.

Lancasters flew 156,192 sorties, twice those of its nearest "competitor," the Handley Page Halifax. Some 152,000 aircrew served in Bomber Command, with 47,268 of them killed in action. The Royal Canadian Air Force lost 9,919 men in Bomber Command. Of 7,377 Lancasters built in the UK and Canada during the war, only 17 remain more or less complete, around the world.

Andrew Mynarski

Scramble! William Lidstone "Willie" McKnight, of Calgary and Edmonton, was a Canadian ace who went through this drill many times in his short career. *Photo: Courtesy Airforce Magazine*

worked so hard to save this piece of aviation history.

CANADA'S FIGHTER-PILOT REPUTATION UPHELD

BETWEEN THE WARS, the RCAF had continued to train some pilots each year, but due to meagre air force financing, few of them were allowed to join the RCAF. Most of those newly-trained pilots satisfied their need to fly by going to Britain, usually at their own expense (or by the between-wars tradition of working across on a cattle boat), and joining the RAF, and soon there were more of them flying RAF planes than there were pilots in the RCAF—400 in the RAF, 235 in the RCAF, in late 1939.

Once the air battles started, many of these expatriate warriors soon upheld Canada's reputation of providing top-of-the-line fighter pilots, young men such as William Lidstone "Willie" McKnight, of Calgary, an aspiring doctor, who became a top-scoring ace in the early stages of war. McKnight fought as a member of RAF 242 Squadron, consisting mainly of Canadians and conceived as a propaganda instrument. A squadron of Canadians—particularly with some French-Canadians—serving in France would reassure the French that Canada would soon be sending forces to help them. Canada had been adamant it would not send Canadian fighter squadrons to Britain, busy as it was preparing for home defence and launching of the huge commonwealth air training plan.

McKnight joined 242 Squadron on Nov. 6, 1940, as an Acting Pilot Officer on a short-service commission. Most of the pilots had earned private flying licences at home, and were from British Columbia, Alberta, Saskatchewan, Manitoba, and a couple from Ontario, comprising such backgrounds as a student of music, a physical fitness instructor, former RCMP constable, salmon fisherman, bank clerk, hardware merchant, hockey player, and band leader. McKnight was considered a character at school in Calgary, and showed a rebellious streak. In high school, he quarterbacked the football team. He had been a so-so medical student, and hoped to get on with

JIM SHILLIDAY | 91

his career after the war by studying at Edinburgh. While training, he was undisciplined, being confined to his barracks twice and, along with a classmate, being placed on open arrest as "perpetrators of a riot." But he settled down when Germany invaded Poland and the recruits were graduated as fighter pilots

By December, the Canadian airmen, most of whom had no operational training, had been given Hurricanes to fly, McKnight getting his first Hurrie flight in February. The following month, he flew his first operational sorties, convoy patrols over the Channel. By May, he was operational for night flying.

On May 10, Germany attacked Holland, Belgium, Luxembourg and France. The RAF squadrons on the Continent took heavy casualties. On May 14, McKnight and nine other pilots with no combat experience, went into action in France attached to RAF units already fighting the invaders. His first day in France, McKnight flew two patrols. From May 14 to May 19, he flew many sorties, shooting down his first aircraft, an Me-109. On one flight, small-arms fire holed his gas tanks. Then the RAF pilots returned to England, having downed six of the enemy during their first combat duty. Despite France's calls for more RAF squadrons, Churchill reluctantly decided to hold them all for defence of Britain.

On May 27, the evacuation of Allied troops from Dunkirk began. The squadron flew out of French airfields during daylight hours, returning in the evening to RAF Manston, perched at the Channel's edge. In fierce fighting over Dunkirk, McKnight claimed a victory over an Me-109 on May 28, but his Hurricane was hit, and he had to limp home with his oil and coolant systems shot up. Over Dunkirk, in four days, he shot down six enemy aircraft, eventually claiming 10 victories by June 7.

The summer and autumn of 1940 saw McKnight and 242 Squadron stationed at Coltishall, in Norfolk, commanded by Douglas Bader, the legless RAF ace. Bader took a shine to McKnight, and chose him as a wingman. On August 30, the squadron relocated south to RAF Duxford and met a large formation of He-111s with Me-109 and Me-110 escorts. Bader scored two victories while McKnight scored a "hat trick," claiming two enemy Me-110s and a He-111 bomber. The RAF Chief of the Air Staff signalled: "Magnificent fighting. You are well on top of the enemy and obviously the fine Canadian traditions of the last war are safe in your hands." McKnight, the squadron's best shot, was promoted to Flying Officer and awarded the DFC on August 30, 1940, and a Bar in September.

While strafing an E-boat in the English Channel on Jan. 12, 1941, McKnight and P/O M.K. Brown broke off after encountering fire from anti-aircraft guns on the French coast, just as an Me-109 attacked. Brown made it back home but McKnight was lost in the channel, believed shot down by the Me-109. The first Canadian ace of the war, he still was Canada's fifth-highest scoring fighter pilot at war's end, four years after his death. He scored 17 victories, as well as two shared and three unconfirmed kills. He is commemorated on a wall of the Calgary International Airport, and at Hamilton's Canadian Warplane Museum, along with Canada's other air war standouts.

DEATH BY REVISION

MANY PEOPLE RECALL the shameful 1992 CBC TV series, *The Valour and the Horror*. The second episode, Death by Moonlight: Bomber Command, attempted to rewrite history. It trivialized the deaths of so many young men. It had been promoted as a program that would pay homage to the courage and patriotism of young Canadians who served in the war. Instead, it stated that the area-bombing policy was immoral, suggesting the young aircrews were, too. It resulted in aircrew veterans from Canada, Australia, New Zealand, United Kingdom and United States suing the CBC, without satisfaction.

Canada's 15 squadrons within the RAF lost 9,919 aircrew members, a fatality rate higher than any other allied combat unit. But their heroism and sacrifices are seldom honoured in national parades or celebrations. In June, 2008, a motion passed in the Canadian Senate urging the federal government "to recognize service of Bomber Command in liberation of Europe during World War II." Sen. Michael Meighen (Cons.) introduced the motion, supported by Sen. Hugh Segal (Cons.) and Sen. Joseph Day (Lib.).

RESPECT FOR MILITARY HISTORY AND TRADITIONS

WATCHING A MOVING 2000 Memorial Day ceremony for the Unknown Soldier in Ottawa took me back to the grounds of Gordon Bell High School in Winnipeg, watching as my big brother and three of his pals stood astride their bicycles, excited and laughing. It was 1943. They had just joined the air force.

They were tanned, healthy and looking forward to their adventure, fully aware, after almost four years of bloody war, of what they faced. Already, the wall in the main hall of the high school was covered with pictures of boys who had been killed. Bobby, my brother, his nose and cheeks freckled, was 17 years old. It wasn't just the appeal of wearing a uniform; air cadets had been compulsory for years, and he was used to the blue uniform, the insignia and badges. No, these boys were following example: when you got to this ripe age, you volunteered to serve your country, Canada.

Soon there were letters coming to our house every few days, often with snapshots, from different parts of the country during stages of his training as an air gunner. He was 18 when he graduated, rated an expert in the use of a machine gun mounted on a bomber. He was posted overseas to Britain. Home on embarkation leave, he waited for a taxi at our Oxford Street home to take him to the CNR Station at Main and Broadway. The taxi hadn't shown up. Panic. Our dad, a locomotive engineer, phoned the stationmaster and shouted as though he owned the Canadian National Railway: "Hold Number Two-eighteen! My boy's going overseas and he has to catch it!" Then we ran down Oxford to Academy Road, caught the streetcar, transferred at Broadway and…they had actually held the train.

Bobby was stationed at Worksop, a bomber station near Lincoln. He was a flight sergeant now, and mid-upper gunner on a mighty Lancaster bomber. He wrote a couple of letters a week, still including snapshots, but he looked pale, bags under his eyes. He was under strain. He never mentioned the flights over occupied Europe; he talked about good times we had had. He talked about good times we would have fishing and, particularly, hunting, now that he was an expert with guns. He told me that I should not be playing high school football because I was too slight.

On his twelfth mission, in January, 1945—now flying out of Scampton for a large bombing raid far into Germany on a synthetic oil installation near Zeitz, south of Berlin—his bomber failed to return. I got home one evening and met a telegraph boy at the door. He handed me the message that my brother was missing. After a while, we received official notice that F/S Robert Charles Shilliday, R263945, was presumed dead, just a few months before the war ended. He was 19 years old. Bob had left some clothes at home, a couple of sports jackets, shirts, ties. I started wearing them to school. I had the feeling that if I wore his clothes, something of him was still alive.

My mother received five medals her son had earned, and a special silver star minted for mothers of the fallen. But that was it. We didn't know what had happened, where it had happened, where his remains were. There's a terrible vacuum in those circumstances, a void that must have been intolerable for mothers to bear.

Then, in the early 1970s, a wonderful thing happened. We discovered that some years earlier, a Manitoba lake had been named after my brother by the Canadian Permanent Committee on Geographical Names. Shilliday Lake was in Duck Mountain Provincial Park, just south of East Blue Lake, and easily accessible. Now, there was something to fasten onto, a permanent memorial to a dead son and brother. Shilliday Lake.

PREACHING AT THE CANADIAN WAR MUSEUM

ARE WE CANADIANS A CONFUSED bunch? It seems we try to smear our heroes at the same time that we are commemorating their efforts. In the Second World War, the RCAF grew to 250,000 personnel and nearly 100 squadrons; one-fourth of the squadrons under RAF control were Canadian, and all of them were risking their lives.

The Canadian War Museum in Ottawa is a marvellous creation. Its displays offer a unique record of a country's efforts to defend itself and its allies from a barbaric coterie of fanatics who set out to rule the world for a thousand years, but lasted just

JIM SHILLIDAY | 93

AVIATION AND THE SUPREME COURT

SOME YEARS AFTER THE SECOND WORLD WAR ended, a pilot from the previous world war, now trying to keep a flying business alive, "took on City Hall" and won a victory that had permanent repercussions throughout the Canadian aviation and real estate worlds.

Most Canadians who flew with the Royal Flying Corps were remembered for their exploits with a "joystick" in the cockpit; Konrad "Konnie" Johannesson was remembered for his skill with a hockey stick on the ice. But Johannesson possibly should be most noted for his successful appeal to the Supreme Court of Canada that was of vital import to airport operations, and to the real estate business generally.

In 1917, Johannesson went overseas to the Royal Flying Corps and became a lieutenant flying instructor at an airfield at El Khanka, Egypt. He returned to Winnipeg in 1919, played hockey as a defenceman with the Winnipeg Falcons, winning both the Allan Cup and Canada's first Olympic Hockey Gold Medal, in 1920. From 1929 to 1934 he was manager and chief flying instructor at the Winnipeg Flying Club, and administrator of the city's main airport, Stevenson Field, for about 20 years, seeing it through development from a cow pasture of pioneer-flying days to a more modern installation with runways, commercial hangars and a Trans-Canada Airlines terminal. During the Second World War, he operated Johannesson Flying Service, teaching Icelandic-speaking students to fly so that they could join the RCAF.

After the war, he fought a long legal battle that was finally decided in his favour in 1952, in the Supreme Court of Canada. That decision allowed him to found and operate Rivercrest Airstrip and Seaplane base on the Red River in West St. Paul, just north of Winnipeg. The constitutional court battle joined Johannesson and John A. Wilson (the civil servant who had directed the development of civil and military aviation in Canada) as the figureheads in "the first two cases to involve aviation and the Supreme Court of Canada," as noted by Patrick Fitz Gerald and Brian Johannesson, in CAHS. The Wilson case had established federal jurisdiction in aviation; the Johannesson case confirmed it.

There was a half-mile strip of land along the Red River, in the municipality of West St. Paul that, in 1947, Johannesson decided to buy and use as an airstrip and floatplane facility. The site was close to a suburban housing development. The municipality passed a by-law to block Johannesson's plans. An appeal was made to the Manitoba Court of Appeal, which upheld the municipality. The case went to the Supreme Court in 1951, and its decision supporting Johannesson came the next year.

Most traces of the riverside air facility now are gone. "However, there is a legacy," said the CAHS article. "It wasn't that Konnie won his argument. It was that by so doing he saved others from having to repeat the process. In cases brought before the Supreme Court since that time involving airports, the inevitable encroachment of urban sprawl, and other conflicts of jurisdiction, the precedent of Johannesson v. West St. Paul offered guidance. Countless law students and urban planners considering development around airports know the name Johannesson. It is involved in a manner that would probably have bemused Konnie Johannesson. Nonetheless, in this manner he continues to contribute to Canadian aviation."

A busy day at Konnie Johannesson's Rivercrest Airstrip near Little Britain, just north of Winnipeg.

12 years and left more than 17 million of earth's inhabitants dead, hundreds of cities in ruin. But then this Canadian thing intruded. The people who were entrusted with running the show decided that it was necessary to preach a little, to make people feel guilty, to remind visitors that war can be bad. They displayed an insensitive panel describing the "Enduring Controversy" over how Canadian airmen used their bombers to decimate the Nazi war machine by destroying industry and large cities.

"Mass bomber raids against Germany resulted in vast destruction and heavy loss of life," says the panel. Of course they did, as did the German bombers flying over Britain. This was a fight for survival, which should be realized and not forgotten. "The value and morality of the strategic bomber against Germany remains bitterly contested…Although Bomber Command and American attacks left 600,000 Germans dead, and more than five million homeless, the raids resulted in only small reductions in German war production until late in the war." Propaganda Minister Goebbels would have approved that propaganda.

Understandably, thousands of Second World War vets were indignant at being characterized as "war criminals." The museum people refused to remove the panel, then asked four historians to study and report on the controversy. Despite the opinions of two historians that the offending panel was "unnecessary" and could be removed, the museum president then refused to alter the panel. The historians were also concerned by "tone" and "balance" of the panel and its illustrations, one showing dead civilians on a German street after an allied bombing. The museum president said he was backed by his 13-member board of trustees, which included the museum's director-general, representatives of veterans' organizations, and three from the general public. The president's stubborn refusal to change the display prompted the Royal Canadian Legion to call again for a public boycott of the museum, and to ask for an investigation by the Senate committee on veterans' affairs.

Young Canadians went to war willingly, out of a sense of duty. They were fully aware that death was possible, if not likely. Nazi submarines prowled the Atlantic, right into the Gulf of St. Lawrence. Hitler's hordes, after practising in Spain where they invented mass bombing of cities and towns, took over most of Europe. Then they set their sights higher. When Canadian airmen dropped bombs over the Continent, they were participating in all-out resistance, which is not gentlemanly. It is a phenomenon known as "total war." One of the better commentaries that I have read was that taken by Lt.-Col. Dean C. Black, who had just been named executive-director of the Air Force Association of Canada. He wrote in *Airforce Magazine* that the issue was "who possesses the right to mandate historical interpretations…Veterans seem to prefer a commemorative voice over the subjective musings of an historian." Musings that might fit better in the world-interpretations of the proposed Canadian Museum of Human Rights than in a war museum that should simply commemorate and inform. "The panel suggests that Bomber Command crews fought in vain and in morally questionable ways. Historical evidence now says otherwise," Black wrote. The display seemed to ignore what total war means, that strategic bombing was an aspect of war lacking in purpose. Strategic bombing was the only means by which a second front could be opened, to relieve pressures on the Russian front and to immediately threaten German economic production. "Surprisingly, the panel in question does not commemorate the 10,000 Canadian bomber aircrew who never came home, but suggests instead that allied aircrew acted on a dubious moral plane." My brother—a Lancaster crew-member who didn't return—was not a 19-year-old immoral person.

By 2007, the Canadian Legion had suggested a boycott, and withdrawn an offer of a substantial cash donation from each branch to the museum. Then, that same year, the museum directors abandoned political correctness and their attempt to rewrite history and installed a new panel that appeared to end the discontent of veterans and many other Canadians. New wording concerning the Second World War strategic bombing campaign of Germany brought the whole issue into a broader context and perspective. Now it was "respectful to history and to veterans." •

No.1 Fighter Wing Sabres put on a power show above North Luffenham's main gate, circa 1953. Faintly drifting behind are exhaust trails, "at three dollars a yard".

| 96 | A MEMORY OF SKY

CHAPTER SIX

THE COLD WAR

ON A SUNNY JULY MORNING in 1953, my instructor, Flying Officer Donald Sweetman, led me out to a breath-taking flight line of streamlined machines, T-33 trainers, and we climbed into No.686. We flew for one hour, a familiarization flight. We flew nine more times over the next few days, formation flying, practice flame-outs, aerobatics, simulated instrument flight, Automatic Direction Finding and Ground Controlled Approaches, and "circuits and bumps" (practise landings and overshoots).

The plane handled beautifully. Formation flying, with its constant minor adjustments for catching up and slowing down, was where the transition from piston to jet engines was most apparent, particularly when joining up. In the old prop-driven Harvard, the 650 horsepower Pratt and Whitney provided almost instant throttle response. The 5,100 lb. thrust of the Rolls-Royce Nene 10 in the T-33 experienced a lag in throttle response, and if a pilot announced "coming on board" and didn't anticipate soon enough, he could scatter the others up, down and sideways. (Just as with the Sabre, the Canadian-built T-33s had better engines than the American. Ivan Henry, a fellow instructor at Gimli, piled up 4,073 hours in the Canadian-built T-33 and 318 in the American-built. "The Nene 10 engine in the Canadian gave our version much better performance than the GE engine with its 4,600 pounds of thrust."). Rolling the Yellow Hazard meant some adroit conjunction of rudder pedals and stick work; to roll the T-Bird, you pulled up the nose and twisted her around.

Sweetman, a quiet personality, had an effective way of making a lesson stick: on my fourth trip, piloting blind under the front-seat canvas hood watching the altimeter unwind during a descent, we passed

YEAR 52	AIRCRAFT		PILOT, OR 1ST PILOT	2ND PILOT, PUPIL OR PASSENGER	DUTY (INCLUDING RESULTS AND REMARKS)
DATE	Type	No.			
—	—	—	—	—	TOTALS BROUGHT FORWARD
			COMMENCED FLYING TRAINING		
5	Harvard	831	F/O Cruickshank	F/C Shilliday	Familiarisation
					TOTAL WEEK-ENDING
			George R. Ayre F/O OC "E" FLT		COURSE TOTAL
					GRAND TOTAL
11	Harvard	2684	F/O Hutchinson	Self	L.P.2
					TOTAL WEEK-ENDING
			George R. Ayre F/O OC "E" FLT		COURSE TOTAL
					GRAND TOTAL
					TOTAL WEEK-ENDING
					COURSE TOTAL
					GRAND TOTAL
17	Harvard	831	F/O Pickles	Self	L.P.3
19	Harvard	940	F/O Pickles	Self	L.P.4
					TOTAL WEEK-ENDING
			signature OC "E" FLT		COURSE TOTAL
					GRAND TOTAL

GRAND TOTAL [Cols. (1) to (10)]
4 Hrs. **55** Mins.

TOTALS CARRIED FORWARD

through ten thousand feet (I thought) and suddenly the hood was released. My heart leapt—we were headed straight into a New Brunswick evergreen forest. "I have control," the instructor growled, taking the stick. I never misread an altimeter again.

The tenth flight was my solo check. However, my first jet solo was not in the T-33. After we had left the trainer and debriefed, I (with just over 13 hours of dual jet experience) was sent out to solo in the front-line Sabre jet fighter, a plane I had sat in but never lifted off the runway. We didn't know there was any other way to do it.

This first flight in a Sabre did not begin auspiciously. After taxiing to the end of the runway and, using the unfamiliar nose wheel steering—which involved engaging a button on the control column and using the rudder pedals—I attempted to line up for take-off. My efforts succeeded in "cocking" the nose wheel so that the plane was pointed appropriately down the runway, but the nose wheel had turned sideways, and wouldn't budge. My first call to the tower from a Sabre wasn't for take-off clearance, but for help. Ground crew came out shaking their heads; aware of their proximity to that sucking intake, one put his back under the nose got it nodding up and down, while the other manhandled the nose wheel into position, allowing me finally to take off...with my tail between my legs. But the rest of that thrilling trip made up for the shaky beginning.

A little more time was logged in the T-33 at RCAF bases in England, Germany and France, the most concentrated period being a two-week instrument flying course at Zweibrucken to earn a "green ticket" instrument rating. Instrument flying was always particularly enjoyable for me, and the T-33 was a steady bird for that.

Every flight in an airplane was exciting, but some more than others. Example: on May 23, 1956, Ray Oldfin, our flight commander and instrument-flying check pilot in Marville, France, signed us out in No.179 for what was to be a ninety-minute flip. Taxiing out, the smell of aviation fuel in the front cockpit was strong, but my comments elicited only an "Mmmm!" from the back seat. There wasn't much bare flesh between my crash helmet and the white silk scarf around my neck, but a dampness seemed to be spraying between my ears from the heat and vent system. I removed my brown kid glove, rubbed a palm on the back of my neck and checked. My hand glistened. Then, the sniff test.

Aviation fuel! But Oldfin was unconvinced by my warning. So I released and raised the cockpit canopy and suggested that he stop, cut the engine and disembark. Still unimpressed by my weak-kneed complaining, he agreed to turn around and taxi back to the tarmac. It turned out there was a fuel leak in the plenum chamber, and the cockpit, indeed, had been filling up with fine droplets of fuel. A bomb!

After three years flying Sabres in Europe, it was back to Canada and Trenton for the instructors' course, my most concentrated experience on T-33s. Then, on to advanced flying school at Gimli, Manitoba, to instruct students from Canada, Britain, Turkey, West Germany, Netherlands and Greece.

Expiration of my RCAF commission sent me back to my true vocation, newspaper work. But the resonant, remembered time of my life, the sublimity of that other world above, was flying jets – the ultimate in escapism, pure and easy and smooth and relatively quiet, free of the shaking and mindless howling of the reciprocating engine. It was the T-33 trainer that prepared us for the superb Sabre (what some now call a "real" airplane), the Spitfire of the jet age, the last pure eyeball (non-computerized) fighter, our pride and joy. We can't be sure that our particular efforts in Sabres did a lot to keep the world free. But all of us (there still are several hundred members of SPAADS, the Sabre Pilots Association of Air Division Squadrons) were ready to try.

AUSTERITY BECOMES PROSPERITY

IF THE VIMY RIDGE BATTLE of 1917 was Canada's coming-of-age event, the country "grew up" during the Second World War. By the end of the Second World War, Canada had the world's fourth largest air force, had built 16,000 aircraft (in the 20 years after the First World War, just 678 aircraft were built in Canada), mostly for Britain, including swarms of Hurricane fighters, Mosquitoes, Lancaster and Halifax bombers; assorted trainers such as Harvards,

Ansons and Tiger Moths that were bound for the BCATP.

Sixty percent of British soldiers carried Canadian-made small arms, such as rifles, Bren guns, anti tank weapons, and millions of rounds of rifle, machine gun and pistol ammunition, 75mm tank gun shells, 25-pounder artillery shells, naval torpedos and anti-submarine depth charges shipped overseas from Canadian manufacturing plants, mainly in Ontario and Quebec. The nation had shaken off the misery of the Depression years, employment was high, industry was thriving, former service personnel were being absorbed by the civilian work force, and women had acquired a measure of liberation. Austerity had become prosperity. Life was good!

With relief, Canadian authorities shed the reminders of war. "Many Hurricanes, Ansons, Harvards and Swordfish ended up on farms across Canada…You could buy a Fairey Battle for $35, a Hurricane for $50, and a Mustang for $500. But, for lack of a market, many were simply crushed by bulldozers, or burned," wrote G.R. Guillet, a former Vampire pilot, in *CAHS*. Then, the defence department showed it was really on the ball: "But the ongoing unease about Soviet intentions resulted in the RCAF restarting pilot training in the fall of 1947 at Centralia…It was decided ultimately to re-equip with Vampire F.3s and P-51 Mustangs." The Vampire, which missed the Second World War by a few months, probably would have dominated enemy fighters of that time. In the late '40s, newly-trained pilots in their Vampires were nicely prepared for the introduction of the Sabre jet fighter.

The air force's built-in training know-how, though vastly scaled down, still was the basis of Canada's flying program that carried over to the early post-war years when the country geared up, with a sense of urgency, for a commitment to NATO service. Most of the instructors were graduates of the BCATP, and RCAF leaders had entrenched dedication to a high standard of training by their own pilots and technicians. There was competence inherent in controlled service teaching that produced top-calibre graduates and a high degree of professional pride with a made-in-Canada stamp.

This was the training standard that pilot trainees of the 1950s experienced. These were the instructors who took us up in Harvards and T-33s and turned us into the best Sabre fighter pilots of the time.

THE SHIFT TO COLD WAR

THE COLD WAR CONFRONTATION between East and West was a fission-folly balance of power that pre-occupied the minds and defence spending of many nations for almost half of the century of flight that began in the early 1900s.

In the 1950s, North American children were being told to hide under their classroom desks in the event of a nuclear attack, and citizens became used to hearing terms such as Strontium-90, concrete bunkers, fall-out, and emergency evacuation routes.

A gigantic Canada-U.S. mutual-defence building program was established between 1952 and 1958 to set up three lines of defence for North America from attack over the North Pole.

- The first line of defence against attack was the Distant Early Warning (DEW) Line, consisting of 58 sites (FPS-19 search radar) at roughly 100-mile intervals, built between 1955 and 1957 along the 70th parallel in the Far North.
- The Mid-Canada Line was the secondary detection that went into operation in 1958, consisting of 90 unmanned sites about 30 miles apart along the 55th parallel, with eight sector control stations.
- The Pinetree Line was composed of 44 long-range radar stations, and six USAF manned Gap Filler radar stations along the 50th parallel. Some of these locations remained operational for more than 35 years.

CF-100 Canuck all-weather fighters and Bomarc missiles (a politically-bungled program in Ontario and Quebec that was supposed to replace the manned Avro Arrow project) backed up the radar sites, and later, CF-101 Voodoos. Just south of the international boundary, a continent-wide stretch of Anti Ballistic Missile nests defended against northern attack. It was never mentioned to Canadians, and few realized, that the American missiles, if launched, would explode their nuclear warheads above Canadians heads.

FLYING TRAINING IN THE T-33

THE SPIRIT AND MEMORY of the T-33 trainer will last a long time. There's hardly a major runway anywhere in Canada that hasn't supported one. Thousands of pilots across the country experienced its delights. And lovingly-preserved specimens banking on pedestals in dozens of cities and towns and airfields will remind Canadians for decades to come of an aviation marvel.

My jet-flying career started and ended with the T-33. The first climb into the cockpit of a T-Bird came in 1953, unaware that it would become one of the aviation workhorses of the century and a true and loyal friend of Canada's air force for the next 49 years, breaking the long-service record of the magnificent DC-3 "Gooney Bird" transport. After investing 280 hours learning to fly the tricky prop-driven Harvard with a semblance of expertise, it was like climbing from a popcorn machine into an electric blender.

Thousands of young Canadians learned audacity while flying this reliable jet trainer. In that other world above, freed from earth's limitations, they acquired self-reliance. It's difficult to think of another trainer so balanced, yin with yang. Unobstructed visibility from its canopy allowed pilots full appreciation of sky, the air absolutely clear but for wispy cirrus high above and daintily stretching dozens and dozens of miles to the horizon. Those who preferred the challenge of dirtier flying, the towering cumulo-nimbus storm clouds churning, their black menace demanding precise instrument competence for safe penetration back to earth—especially when there was someone on their wing—couldn't fault the steadiness of the ubiquitous "T-Bird."

Harry Bauer wrote in The Flying Mystique: "Because it is our own judgment at work, our own decision to be in the air and rely on ourselves, we are for the time we are flying living life at its fullest." He certainly was writing about all those Canadians in their T-33s.

The T-Bird ended its career with the RCAF and the Canadian Forces in April, 2002. It began life as the P-80 Shooting Star (the United States' first operational jet fighter) then was stretched, fitted with an additional seat and a huge canopy, renamed the T-33 (Silver Star in Canada) and built under licence by Canadair. By 1953, RCAF and NATO pilot trainees were taking advanced flying training in the now beautiful trainer at 3 Advanced Flying School Gimli and, soon after, at 2 AFS Portage la Prairie and 1 Pilot Weapons School, MacDonald, all in Manitoba. So those subsequently posted to F-86 Sabres, or CF-100 Canucks, had lots of jet experience.

But my intake, 29, missed this early jet-flying experience. The T-33s hadn't arrived yet and we earned our wings at Portage flying the propeller-driven "Yellow Hazards," and then finished Harvard trainer days at MacDonald, firing .303 machine guns, wobbly rockets and dropping tiny powderless bombs at the Langruth range, on the western shore of Lake Manitoba. Innocent of any experience flying jets, we were sent on to No.1 (Fighter) Operational Training Unit at Chatham, New Brunswick, to become proficient F-86 fighter pilots.

JIM SHILLIDAY

⇒

An RCAF instructor of NATO students, Flying Officer Jim Shilliday climbs into a T-33 Trainer at 3AFS Gimli, in 1957.

Canadians were exposed to endless inter-bloc chess-playing, and near-war challenges, but heard little of the flying activities of their airmen, and recall almost nothing of the role of their air force during the critical years of the 1950s. The invasion of South Korea by Communist hordes in 1950 was interpreted as part of a larger plan of domination, and Canada snapped to attention in its new NATO commitment. It had Sabre fighter squadrons taking up defence positions in England by 1951 and on the continent soon after. This "peacetime" defence spending was the biggest outlay since the Second World War.

What was the Cold War? What were we doing flying fighter jets in Europe? Russia, just four years after what it called the Great Patriotic War of 1941-45, had perfected an atomic device, and then a thermo-nuclear bomb under Stalin's "science of war" doctrine (never before in military history, anywhere, had the methods of killing people changed so much and so quickly). Up to then, it was believed, the Russians would have hesitated to initiate any local military action for fear of nuclear retaliation. But once they had nuclear equality, they were prepared to risk military incursions. The RAF's Air Chief Marshal, Sir John Slessor, called this "termite tactics." The other side would try to gradually extend the Communist empire by local military-political action.

There had been attacks on NATO aircraft in Europe. The U.S.S.R.'s military policy was what war planners called "mischief short of war." But some strategists had feared that what had happened in Korea could be repeated in divided Germany. There were scores of advanced tactical air bases in central Europe controlled by the Russians. Soon thousands of MiG-15 fighters and assorted bombers were scattered all through these airfields, and MiG-17s were on the way. Today, not much mind is paid to what the RCAF was doing in the mid-1950s. But to the Canadian fighter pilots, there were no certainties, the chance of the unbelievable happening was very real. They trained hard, and were ready to fight hard.

THE RUSSIANS

WAR AND PEACE. EAST AND WEST. Oil and water. Always there are two sorts of truth in opposition. Tolstoy, in 1868, had written his own sort of truth: "The cudgel of the people's war was lifted with all its menacing and majestic might, and caring nothing for good taste and procedure, with dull-witted simplicity but sound judgement it rose and fell, making no distinctions." In the 1950s, that cudgel was threatening to rise again.

Much of Russia's topography is similar to Canada's—flat or undulating, with forests and mountain chains. And, as in Canada, it was a natural place for early aviation to thrive. But Russia has 10 times the population of Canada, and a suspicion of outsiders Canada will never have. When you scratched a Russian, the blood that flowed was sluggish with the hard-learned lessons of history. This blood culture was conditioned by seven centuries of struggle to recover from the ravages of Genghis Khan and his successors, who snuffed the candle of enlightenment and doomed Russia to remain in the dark ages long after peoples to the west were stimulated into new heights of what they called civilized behaviour.

Ever since Ivan the Terrible got rid of the Mongols hundreds of years before, Russian leaders had emphasized defence over an easier life for their people. A choice of more food or more guns meant hardware won, no question. Before Ivan, Russia had been backward in every way where modern nations had progressed—in education, industry, social development, military preparedness. She had been mauled by any strong nation that could get to her. Stalin determined to drag the country even further out of its bog and spite the hostile world he saw around him. "Either we do it or they crush us," he said, in order to justify the ruthless measures he proceeded to implement.

But he was not original. The same draconian steps—labour camps, prisons, barring of foreigners, Siberian exile—all had been practised by the tzars. When Stalin died in 1953, the new Russian rulers heaped blame on him for past failures and oppression. But, of course, they carried on the tradition. Guns before butter. Don't trust the foreigners. Get them before they get us (to a large degree, it's still that way today). This was the inevitability-of-war doctrine. The great union of socialist republics—that lost 28 million people fighting our common Nazi enemy—was developing a nuclear capacity and delivery system and an air defence network that would be able to meet any nuclear attack on the Soviet Union.

So, how militarily strong were the "Reds" in comparison with the West? The military regime was reorganized after Stalin's death with a unified ministry of defence. The training of their flyers was almost the same as that of NATO. The chairman of the U.S. Joint Committee on Atomic Energy stated that Russia in three or four years could launch a saturation attack on the United States. "She possesses all four of the main ingredients of a strong atomic warfare program: adequate material resources, including uranium; adequate scientific competence; adequate technological capacity; adequate production capacity."

Russian pilots were encouraged to discuss all these things over and over. They were thoroughly acquainted with all of their leaders' arguments justifying every military tactic, including a few early incursions into Western air space. They flew those missions without question. They were ants in a colony. The nuclear umbrella was like a huge fallen tree that protected them; the West aimed to stop them from rushing out to raid other nests, to expand their sphere of influence into a hegemony across Europe. In March, 1954, the U.S. was still detonating hydrogen bombs in tests conducted in the Marshall Islands.

Soon, we Canadians were playing cat-and-mouse with MiGs along the buffer zone skirting the west side of East Germany, Hungary, Yugoslavia,

Romania, Bulgaria and Czechoslovakia, now all Communist client states and, in our minds, mean and definitely not Marxists of the Groucho variety.

THE CANADIANS

WE CAME DOWN FROM THE CLOUDs, our heads still full of sky, and marvelled at being Canadian aviators. A holistic feeling, everything right, from all points of the compass.

Our aircraft—the Canadair Sabre—was the best in the mid-twentieth century world; as pilots, we granted no-one superiority; as servicemen, we respected our superiors (our top leaders were flyers first, politicians second); as citizens, we admired our country. We had pride. We were well trained. We were sure, if called upon, that we would put on a good show.

From our fellow flight cadets, we had learned appreciation of the diversity of our huge country. As we learned our craft, our eyes were opened to the myriad aspects of the land: Flying Training School night flights from the reflections off Lake Huron to the sparkling lights of Orangeville and London; the flat vastness of the Portage la Prairie agricultural belt where sentinel grain elevators ranked as dependable proxy navigational aids at Advanced Flying School; the soggy marshlands of western Lake Manitoba where the rat-tat-tat of a Harvard's single machine gun ruffled the silence of MacDonald Pilot Weapons School's firing range at Langruth; the ruggedness of maritime land and seascapes, from endless, undulating ranges of spruce forest to ragged shorelines, and the majesty of the seaward flow of the Miramichi.

We felt new spirit and took it across the Atlantic with us, to the skies over England and the Channel, where the chatter of earlier fighter pilots still echoed in the mind; along Sabre Alley, that indistinct battlefield-of-the-air west of the Rhine River and more-or-less running along the West German-French demarcation, from

RAF defenders during combat exercises did their best in two-engine Meteors (Meatboxes) and single-engine, twin-boomed Vampires. It was difficult for Sabre pilots not to feel superior.

north of the Swiss border up to north of Wiesbaden, we fought mock battles. If you say we were like children playing, you must remember that children take their play very seriously.

We mock-fought RAF Meteors, Vampires, Venoms, Swifts and Hunters; we fought French Mysteres and Ouragans; Belgian and Dutch Thunderjets; American Sabres and F-100 SuperSabres. We fought, sometimes, in maelstroms of up to three dozen aircraft, where fighters could be nose-to-nose and closing at more than 1,000 mph, with a few feet to spare. We might be out-turned, but using hit-and-run tactics, we seldom were vanquished.

Our generation expected to work hard, possibly because it came from stock that had survived economic depression and two global wars. The world ahead held promise; all things were possible. Our country's aeronautical expertise was outstanding.

We were Canadian. We were the best.

PREPARING FOR WAR

WHEN WESTERN EUROPEAN COUNTRIES were attacked in 1914, young men from cities, towns and farms in Canada joined the fray in their flimsy Sopwith Pups and Camels like wasps erupting. Bishop, Barker, Brown, MacLeod, Collishaw, Stevenson were some of the fliers who saw adventure in air war above the Western Front during the Great War of 1914-1918.

The Second World War saw a new generation, indomitable through the Battle of France, with "The Few" in the Battle of Britain, Malta and D-Day; men such as Beurling, Turner, McKnight, Edwards, Laubman, Klersy, Woodward.

Then, in the 1950s, Britain and the Continent again were threatened with war, and to counter the Warsaw Pact's growing threat, NATO was set up, consisting of Canada, Denmark, United Kingdom, Iceland, Portugal, United States, France, Belgium, Netherlands, Italy, Norway and Luxembourg. The Canadians came back once more, this time in their own swept-wing F-86 Sabre jets of the Royal Canadian Air Force. They fired no shots in anger, but 107 of them were killed, and they certainly were honed and ready to scrap if the green light had flashed.

National delusions lead to disillusionment. It was a fairly recent Canadian characteristic to believe we could help maintain peace by having only peaceful intentions. But Canada's military tradition cannot be that easily re-written. After several earlier wars, Canada was one of the major combatants in the Second World War with a million young men in uniform (losing more than 10,000 sons in the air war alone), ending up

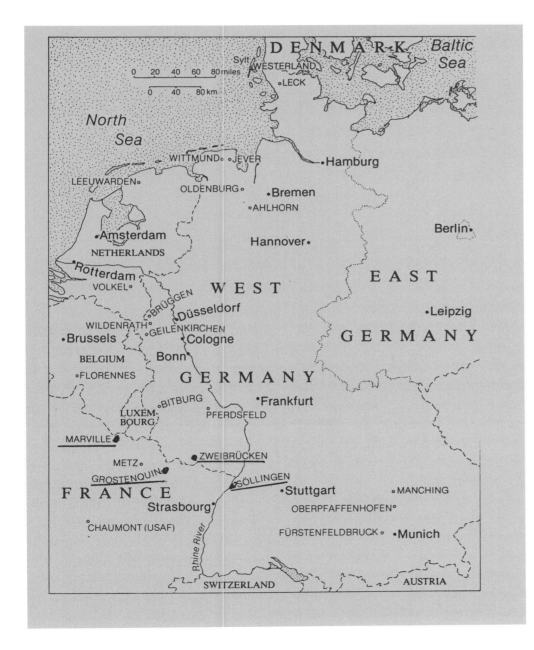

West and East Germany were divided by a no-fly buffer zone. The four Canadian bases are underlined. *Ralph Clint/CANAV Books Collection*

BIRTH OF A RUSSIAN BOMBER

BRIGADIER-GENERAL VLADIMIR Konstantinovich Kokkinaki, of "Moscow to Miscou" fame, helped develop the "Beagle", one of the warplanes that we NATO fighter pilots were to defend against. The early Beagle bomber that we were training to shoot down was powered by engines provided by the British, as were some of the MiG-15 fighters that would try to shoot us down, and had been flown by Communist forces during the recent Korean War.

When the Second World War ended, the U.S.S.R. was way behind the West in designing good jet engines. German engines had been used in early Russian jet fighter designs, but Stalin warned that all means possible must be employed to catch up and produce good home-designed power plants. Fighters were necessary for air superiority so they were the priority. But jet bomber design soon was humming as well, with designers Sukhoi, Tupolev and Ilyushin coming up with the first generation.

Although Stalin wanted "home grown" jet engines, aviation minister Mikhail Khrunichev and aircraft designer Alexander Sergeyevich Yakovlev asked him, according to an article on aircraft designer Artem Mikoyan, in Wikipedia, if the great leader would consider letting them ask the British to sell their turbojet technology. "What fool will sell us his secrets?" he replied. But he said okay and, to everyone's surprise, the British said okay, too, offering the Rolls-Royce Derwent and Nene centrifugal-flow turbojets. So, for a time, the Beagle was powered by British-built Nene engines (as was our T-33) until Russians were making them under a licensing deal. A second prototype flew about six months after the first, being pushed along by now Soviet-built engines.

The Nene provided the Soviet Union with a reliable, powerful turbojet, which went into production as the Klimov RD-45F. Full production of the Il-28 Beagle began in 1949 (the year NATO came into being), and on May Day the next year, appeared over Moscow's Red Square to show off the state's new bomber. Factories were supplying Red Air Force squadrons by March, 1950.

with the world's third-largest navy, fourth largest air force. This country was the training school for the British Empire's fighting airmen, said to be its greatest contribution to the war effort.

More than 27,000 Canadians fought, and 517 died, in Korea, the conflict that spurred Canada into the Cold War—and during that East-West confrontation of the 1950s, mainly in Europe, Canada had an outstanding record of ungrudging contribution. For a brief, proud time it had the best aircraft and pilots of any air force in the world, at a time when both sides refused to hold high-level talks, disseminated intense propaganda, and each built more and more nuclear weapons. "More than most people," wrote Desmond Morton in *A Military History of Canada*, "Canadians have distanced themselves from thoughts of war. On the whole, they have no affection for militarism or a military cast of mind. Canada's wars have exposed the fault lines which the Fathers of Confederation had hoped to bury and which successful politicians have always had to patch…Yet, in important ways, Canada's armed forces and Canada's wars have fostered a sense of national identity and pride…they no longer wondered what it meant to be a Canadian."

There were four Canadian operational Sabre wings on the Continent in the 1950s, with an authorized strength of 300 aircraft. Canada shared with the United States and France the responsibility for patrolling the central sector of West European skies, on the alert for a surprise attack from any aggressor. The Canadians straddled the route invaders from the east would take. The wings were controlled from a picturesque old chateau, No.1 Air Division headquarters, five miles from Metz, France (the hell-hole of trench warfare during the First World War), and supplied by 30 Air Materiel Base, at Langar, England, not far from Nottingham. That base—and Canada's fleets of transport aircraft—provided logistical support to all of the Royal Canadian Air Force's fighter bases in Europe.

The first Canadian fighter base serving NATO in Europe was No.1 (Fighter) Wing at North Luffenham, England, set up in 1951. The airfield at Grostenquin, France (Two Wing), was the first of the RCAF's continental NATO bases to be activated. The second was at Zweibrucken, West Germany (Three Wing), and the third, at Baden-Soellingen, West Germany (Four Wing), whose aircraft nestled amongst the fir trees of the Black Forest. Finally, in 1955, the four Canadian wings were consolidated on the continent with the move of No.1 (F) Wing from North Luffenham to Marville, France. As it turned out, and unrealized by us at the time, that move to France was the next-to-final cut to the military umbilical cord joining Canada and Britain, until 1968, when unification of Canada's armed forces "cleansed" the air force of any apparent relationship with the RAF.

The following Canadian squadrons equipped with the F-86 Sabre saw service in Europe between 1951 and 1963: 410, 439, 441, 416, 421, 430, 413, 427, 434, 414, 422 and 444. No. 410 (Fighter) Squadron was the first to arrive in Europe, beginning operations in England in 1951. It was the first to leave England for Marville, France, but because the French-built runways were not ready, spent six months flying out of No.4 (F) Wing at Baden-Soellingen. In those days it would have been inconceivable to depend on moving troops or equipment by rented air transport, as the military has been forced to do in recent years. The move was almost entirely by the air force's own transport, supplied by 426 Transport Squadron. No. 441 Squadron settled in for a few months at Zweibrucken. No. 439 Squadron moved from England directly to Marville. Heavy equipment was flown into Marville with the help of C-119 Flying Boxcars of 436 Transport Squadron, based in

Air Division's No. 1 Fighter Wing at Marville, France, was a kind of plug in the middle of an invasion route that had been used countless times over centuries. Here, the author patrols the skies of northeast France and Belgium. *Photo: Wilf Thorne*

A "six-pack" of Cougars outside their wartime wooden flight shack, 1954. Front row: Steve Atherton, "Tweedie" Vaesen, Ray Oldfin; back row: Cal Drake, Pat Mepham, Bruce Fleming (Korean vet with two "kills" and a DFC).

PULLING THE BEAR'S EARS

WE GOT THE CLOSEST TO OUR ultimate job—defending Western Europe from attacks by Eastern Bloc aircraft—when we flew Zulu, which meant "on Alert", a condition of readiness, not pretending. Always, at least two Canadian squadrons were on Alert, pilots close to their fully-armed aircraft by the runway, ready to take-off in seconds. On most routine training missions, our fangs were pulled. T-bars inserted in the gun mechanism prevented our six Point Fives from being fired. But on Zulu sorties, we were capable of asserting ourselves. About once a month we went into Alert mode at one of our four bases, in France and West Germany.

Usually, our Alert flights were a cat-and-mouse game with Communist air controllers and pilots. Both sides were prohibited from entering a no-fly area known as the "buffer zone." Sometimes we flew parallel with the buffer zone to see what the enemy would do. Most often, we would fly due east, straight at them,

Dorval, Quebec (now Pierre Elliott Trudeau Airport), and 435 Transport Squadron based in Edmonton, Alberta. By April of 1955, the shattering roar of Sabre aircraft was an everyday sound in the Marville area.

Canada also had air controllers among the best on the continent, a small but highly expert contingent whose call sign was Yellowjack. This was No. 61 Aircraft Control and Warning Squadron, a radar unit based about a mile from the chateau housing Air Division's headquarters at Metz. Every Canadian pilot had complete faith in the Yellowjack controllers. It was like talking to Mom. They depended on Yellowjack to get them there—sometimes so close to Russian-made MiGs that they could smell the garlic—and to get them back again. Soon, Yellowjack's reputation for excellence was continent-wide and pilots from all air forces knew they could confidently call the Canadian controllers when needing help.

⋙

Just six years after the Second World War, Canadian fighter pilots were back in England, swarming from base, criss-crossing the Midlands looking for action. Soon, requests piled up for sonic booms at British air shows.

at around 40,000 feet, our radar controllers at Yellowjack watching to see what airfield enemy "blips" would scramble from, in what numbers, how quickly they responded, how long it took them to climb, and so on. Yellowjack would keep us informed of the "bandits'" altitude and distance from us, as we swiftly closed together. More often than not, we would end up flying into the zone to the warnings of our air controllers: "Danger of penetration, turn about now, turn about now! Expedite turn!" As we fled back into Western air space, our controller continued to inform us of the MiGs' manoeuvring, and we mentally pulled in our tails as we considered the possibility of armed MiGs coming up our "sixes"—the position of the clock's hand denoting directly to the rear, twelve o'clock being directly ahead.

These were the golden years of the Royal Canadian Air Force's post-war activities, a time when its pilots and Canadair Sabre jets were masters of European skies. And it was a time when Canadian aeronautical engineers were turning out the first all-jet airliner; the Arrow, an all-weather fighter that was a design marvel; and jet engines just as spectacular.

The Cold War of the 1950s was not recognized at the time for what it was—the beginning of the end of innocence, a time when people still believed that good could overcome evil...even that atomic war was survivable. East and West were preparing for a fight. Britain and United States had amassed bombers to deliver nuclear weapons to Eastern Europe. The Communists were building their air defences, amassing hydrogen and atom bombs, and planes able to deliver them.

To help take on the Soviet Union, Canada, with a population then of 14 million, invested more than $5 billion to bolster the concept of trans-Atlantic defence. In today's money, this would be an astonishing 42 cents of every tax dollar—an undertaking of well more than $60 billion.

And Canada began to train a new generation of fighter pilots to fly the planes to defend freedom. Every airman had his own stories to tell; ahead, are a few of mine.

THOSE FIRST TRAINING FLIGHTS

YOUTH IS NOT WASTED ON the young, because it helps them to cope. Put another way, they're too young and too eager to be scared. The young mind knows only a future, where bad things are not likely to happen.

This was 1952. Vincent Massey had just been appointed the first Canadian-born Governor General; the pure jet Boeing B-52 and Avro Vulcan made their first flights; Canadian troops were fighting in Korea (RCAF Sabre pilots, too); the movies *Breaking the Sound Barrier* and *African Queen* were released; and the first commercial all-jet airliner, de Havilland Comet, entered service with the United Kingdom's BOAC, and soon after, with the RCAF.

From all parts of the dominion, bus and train routes led to Personnel Selection Unit (PSU) at Crumlin, on the outskirts of London, Ontario, where we got our haircuts, strutted and bragged to impress the others, were set upon by Corporal Collins, the parade square drill tyrant, and did our best to make the right impression. Officers with clipboards interrogated us (getting surprisingly personal), subjected us to complicated psychological tests, gave us lectures and quizzes on such unexpected subjects as ethics. As flight cadets, we saluted everyone, but no one saluted us. Here, we first heard the term CT'd, "ceased training," a curse that continued to catch up with some unfortunates later in flying training school. The CTs came frequently at PSU, the failures being cleared off the unit quickly because their dejection was immense and distracting.

And it was at Crumlin that the first career-shaping announcements were made. There were more than 100 young men in my intake and all knew that there were three selections open to them: pilot (everyone wanted that), navigator, and radio officer. One morning, all of us were called to the auditorium to be told our fate; on the stage sat several officers who went at it swiftly and unemotionally: "The following will be radio officers." Bated breath until he finished, and your name had not been mentioned. Then, the list of navigators. The suspense was excruciating, but when your name was not on that list, exultation. A pilot, I'll be a pilot! Of the scores initially starting out, only eight became pilots.

About this time, a flight lieutenant (captain) asked me and another recruit to fly with him to Camp Borden (the training centre for Canada's fledgling air force back in 1920). He was the lone pilot, so I beat my fellow trainee to the co-pilot's (second dickey) seat on the right of the transport/utility twin-engine Expeditor, and we were off. Trouble was, when we got to Camp Borden, our pilot landed in the wrong direction, with the wind. We quickly used up the short runway and I was gaping at the scenario of our pilot fighting to bring the situation under control as what appeared to be a small hill or rock formation, or building (I can't recall), at the other end of the runway rapidly advanced to welcome us to Camp Borden.

Our fearless flyer, when he saw there was no advantage to continuing along the runway, decided to leave it, managed to swing the aircraft to starboard, applying great strain to the brakes and undercarriage. We made a right-angle turn, headed between largish trees towards a building containing classrooms. We gained the attention of the hard-working students. Faces appeared at the windows, gawking. Then, a mass fade-out as all ran for the exits and safety. Our Expeditor stopped, two yards from the windows, props still turning. I could read writing on the blackboard inside, something to do with reciprocating engines.

The second flight (fright?) came at No.1 Flying Training School, Centralia, northwest of London, Ontario. The sunny afternoon of June 5, my brain still befuddled from morning ground school classes, the flying officer (lieutenant) instructor (a wartime Typhoon pilot named Pickles) and I walked out to Harvard No. 831 off the controls, just find out what a service training aircraft feels like, look at the surrounding area. Very interesting. Take-offs, I decided, were great, and I was looking forward to the landing. It was highly improbable that we would see one, but a subject we had covered in ground school that morning had been flares, that red meant don't land, green meant you are cleared to land. There were a couple of other flares mentioned as well.

The view from the Harvard was grand as we descended on final approach, about 500 feet up, tipped forward, my front-seat position affording an exclusive view of the runway ahead, getting closer and wider. Then something arced upwards and puffed above our path…a flare from the control tower, it wasn't green and it wasn't red. A kind of orangey, brownish smoke. Then I remembered—something about two aircraft on the same glide path… danger of collision…fly straight and level. My instructor remembered, too, and did all the correct things so that we didn't come into contact with the Harvard trainer directly beneath us. It all was very exciting. I looked forward to more.

As the calendar advanced, there was no disappointment. Every Harvard take-off and every landing satisfied me in some way. First solo came a while after many of the others had done it, but still that first take-off alone was a life-moment.

The days never were dull. Hearing the Pratt and Whitney cut out while inverted at the top of a loop was titillating. Nothing could beat the adventure of climbing vertically until motion ceased, then the tail slide backwards, downwards, followed by the weight of the engine answering the pull of gravity with a snap that had you suddenly descending face down, rather than back down, no

We pretended the modest Harvard was both a fighter and a bomber over Manitoba's Langruth firing range. No. 1 Pilot Weapons School, Macdonald, Man. *Photo: Courtesy Airforce Magazine*

longer flying, just falling. There was a thrill in flying solo in a four-plane formation. And navigation trips around the province. Fine tuning the dits and dahs of the radio range. Overshooting. And the engine—all those horses, the howling power, the variable speed propeller that altered the sound of the Pratt and Whitney, giving you a lion's undulating roar as you prowled the sky.

The first night flight from a small runway near Grand Bend on the Lake Huron shore: full power, tail up, charging from black into black, eyes fastened on occasional dim blue lights that you must not cross left or right, and finally lifting up, eyes now fastening onto dimly-lit cockpit instruments to maintain life. And the stop-start of the heart on glancing out of the cockpit and your Harvard is on fire—no, it's flames from the exhaust. They didn't mention that.

The Harvard's powerful engine meant that the propeller torque tended to twist the aircraft off the runway on landing. This could cause a "ground loop," and to prevent the plane from lurching out of control, intense concentration was a good idea after touching down. My FTS mid-course proficiency check, the Partial Harvard Handling, came when there were 81 hours of flying time in the logbook. This involved going over everything we had learned so far, including general airmanship, aerobatics, instrument flying and cross-country navigation. On the landing approach, feeling everything had gone reasonably well, my mind was relaxed, confident. We rounded out, touched down, and began our landing roll along the runway, still at high speed. Then it happened.

The Harvard began a strong swing to port. I was the pilot in control and immediately applied heavy right brake to stop the swing. Unfortunately, my instructor of little faith slammed his foot onto the right brake at the same time. This was too much for the Harvard and it gave up in disgust, swung violently to starboard instead. Now, although continuing along the runway on course, the port wing led the way, the nose pointing at right angle into the grass alongside the runway. Shaking, banging, bouncing, body straining against the seat straps, dust and grit. Then stunned silence. The instructor's voice in my earphones: "Let's get out before it burns." Push back the coupe top, climb out heavily, still weighed down with parachute, onto the runway, strangely close at hand, then we see the two main wheels and struts back on the runway where they had parted company seeking relative safety on their own. The port wing is curled up, as are the propeller blades. There's a bulge in the coupe top, caused by the instructor's head. We're okay. A landing of the highly unconventional kind.

A car drove up, parked on the runway amongst the pieces and the engineering officer got out to shake his head. We threw our parachutes inside and were driven to the hangar. My instructor went into the flight commander's office to talk things over. I sat down by the windows, stared at the black ties stapled to the wall near the ceiling that had been cut from first-soloists who had sung the joys of flight, and thought about "CT" (ceased training) for the next hour, sinking into depression. Then my instructor strode from the flight commander's office, parachute over his shoulder and shouted: "Let's go Shilliday!" The old saw: when a horse throws you, climb back on before your nerve deserts you. I was still an air force pilot.

Our official rank was flight cadet. But a mock command structure was set up after a poll of students determined which of our colleagues ranked highest in our minds. We wore the shoulder ranks that fitted the poll. Our student "boss," fitted with a Group Captain's (Colonel) epaulets, I believe, was Bernard Williams, a Royal Air Force officer cadet serving his two-year national service, and in Canada under a NATO training agreement. He signed his notice board messages to us and added, "Great White Chief." We heard that he had been a lecturer in philosophy at Oxford.

The RAF trainees, about a dozen of them, were a fine bunch. We mixed well, but there was a certain reserve about them until they got used to the language many Canadians employed. For emphasis, the tendency was to use expletives. It's doubtful we were particularly aware of our foul tongues, but Williams and his pals made their point at lunch one day. Enjoying themselves, they politely asked each other to "Pass the effing butter, please," "I say, would you like some effing bread?" "Don't you think the soup could use some more effing salt?" That simple ploy resulted in a noticeable drop in swearing in the mess.

There was something about Williams that challenged me. When he expounded, I often found myself contradicting, or presenting an alternative. He had no trouble shooting me down. But my manner must have rankled, because one evening in the mess lounge, he confronted me while some of his friends gathered around, smiling expectantly. He wanted to know what I thought about "free will," and other philosophical issues. I had trouble presenting coherent answers. This encounter seemed to bolster our relationship, and when he flew to Winnipeg for a visit, he bunked at my mother's home for a night.

The RAF group returned to Britain following the course at Centralia. I heard no more of Williams. Then, in 2003, while reading the *Globe and Mail*, I noticed an obituary taken from *The Guardian*. The photo of an old man staring out at me I

Flight Cadet Jim Shilliday (left) with unidentified course-mate and two of his RAF friends, Ted Stratton (killed back home in a Meteor crash) and Bernard Williams, Shilliday's nemesis – Centralia, Ont., 1952.

 No. 1 (Fighter) Wing, RCAF Base North Luffenham, "Spearhead of NATO", on old Luffenham Heath in the Midlands, where the runways were poured in the 1930s. The 410 Squadron hangars are the three black ones, centre bottom.

recognized almost instantly—my erstwhile flying buddy Bernard Williams. But this man who had died was known as Sir Bernard Williams. "He was arguably the greatest British philosopher of his era," the obit said, "who revivified moral philosophy…." What transported me back to those Centralia days was the second paragraph: "Dazzlingly quick and devastating in discussion, he was famously able to summarize other people's arguments better than they could themselves, and anticipate an antagonists's objections to his objections—and in turn, his objections to theirs—before they had even finished their sentence…." It also pointed out that two "famous examples" of his work referred to philosophical conundrums involving an individual he called "Jim."

There's a picture in my album of Williams and myself standing by a Harvard, and the old urge to disagree with him surfaced when I read, "Mr. Williams did his national service in the Royal Air Force—the year he spent flying Spitfires in Canada was, he sometimes said, the happiest in his life." But *Spitfires* had to be the obit writer's lack of connection with that time.

Leaving Centralia for Advanced Flying School at Portage la Prairie, Manitoba, we were now pilot officers and had invested in officer's flat caps, best-blue dress uniforms and dashing winter great coats. At Portage, we trained with a group of French cadets and both national anthems were played at our wings parade. Some of these Frenchmen displayed an "esprit" more suitable to Napoleonic times. One of them, Fisse, listened to our discussion of safe procedures for descending through heavy cloud, shrugged and exclaimed: "When I want to go down, I just push stick forward and go down." A few days later, he crashed and was killed, after "going down" through cloud.

After sprouting wings and promotions to Flying Officer at Portage, we shifted a few miles north to No. 1 Pilot Weapons School, Macdonald, just below Lake Manitoba. There, we pretended the modest Harvard was both a fighter and a bomber, expending .303 machine gun bullets, small zig-zaggy rockets and explosiveless bombs at the defenceless Langruth range, just off the west shore of the lake.

Now, the final selection was made. Some of us would fly Sabre jets, the others would fly CF-100 all-weather jets. I was the happiest man in all creation after learning I would take the train to No.1 (Fighter) Operational Training Unit, Chatham, New Brunswick. I was going to be a Sabre pilot. What could be better? And soon I would be in Europe.

LUFFENHAM'S SWORDFIGHTERS

Looking down from the clouds, pilots see that the waters of The Wash shimmer wetly in an attempt to produce a mirror-image of the sun. From this 1954 vantage-point to the haze-welded horizon in all directions is encompassed an area roughly bounding the east Midlands and East Anglia. Fighter pilots are nameless glints in the sun now. They fly faster than anyone in these skies before. Their bounding lightness is circumscribed by just two realities: survival is their own responsibility; their purpose is the survival of others.

This perspective, above the lower half of England, shows that the bay of water known as The Wash is the toothless mouth of a crone, East Anglia forming the gummy lower lip and protruding jaw; the East Midlands the shrunken upper lip. Just a short flight to the east, a hundred miles or so, brings wild, flat beaches and marshlands. Across the bottom of The Wash, across The Fens and to the east is East Anglia, the land of long horizons and The Broads. This is the visor facing the foe that has taken the brunt of many blows since neolithic man roamed the Lincolnshire wolds. Its familiarity with the invader's sword and military presence has marked the region, left its dialects, its place names, its land routes and, most visible of all from this altitude, its airfields.

The latest indelible mark left by military design was laid down between l937 and l945 when scores of airfields were constructed as bases for fighter planes furious to protect their homeland, and as integrated hornet nests that issued winged attackers collecting in the thousand-bomber swarms which flew into occupied Europe and Nazi Germany to attack the menace from the east. Clusters of runways everywhere. The military chart is blue with dots designating Second World War airfields—Witchford, Mildenhall, Lakenheath, Mepal, Waterbeach—few more than ten miles apart, saturating the lower half of England, a colossal, misshapen aircraft carrier anchored just a few miles off the coast of continental Europe. Now, military jet pilots are reasonably sure that if they fall powerless from the skies they can emerge from low clouds and glide to safety on a hard surface of their choice.

From their cockpits, pilots note that The Wash's south coast is an udder with three river-mouth teats, the left one jutting wetly to the southwest on a bearing of about two hundred and twenty-five degrees. Helpfully, it points almost directly towards North Luffenham airfield thirty miles away. Old Luffenham Heath, where the runways were poured in the 1930s, is just a few miles west of The Fens on slightly rising ground giving an altimeter setting for three hundred and fifty feet above sea level. The airfield nestles just south and east of Rutland's centre.

North Luffenham air base fits in unobtrusively, its rough edges and tonal contrasts blunted by years of wartime use and peacetime encroachments. You can see three landing strips down there—the main one extending east and west—contained by an outer perimeter, or taxiing route, clustered with aircraft dispersal areas known as "marguerites." Some of these outer works have reverted to the bucolic—the grazing of sheep, the browns and scarring of farm work.

Dropping like a hernia from the bottom right of the airfield is the support and service area—the barracks, hospital, other-ranks mess halls, chapels, administrative offices and, closer to the runways, the control tower and two huge, brick hangars housing the pilots and support staff of two fighter squadrons, 439 (Sabretooth Tigers) and 441 (Silver Fox). And in prized isolation, far across towards the east side of the field, are three older, tar-domed hangars. Between them and the tarmac, punctuated by a couple of stovepipes, is a war-time shack weathered grey, not much longer than a railway boxcar, a hand-painted sign to the left of the entrance emblazoned with a bare-fanged beast and proclaiming that this is the home of 410 (Fighter) Squadron. This flight shack, this confined, uncomfortable, deficient throwback to an earlier Royal Air Force era is the Cougars' castle, the battlement from which they surge to mount their steeds and ride into conflict, the core of their military sensibilities that subtly adrenalizes, a place always, in real time and memory, that is special.

The Canadian-made Mark 2 Sabres at North Luffenham pushed out 5,200 pounds of thrust on American GE 13 J-47 engines; soon they had

Mk.5s with a Canadian-built Orenda 10 engine and 6,500 pounds thrust. When they moved to France, they would fly Mk.6s, an Orenda 14 with 7,500 pounds, an aircraft superior to any American-built Sabre and one that even the great American fighter pilot Chuck Yeager (first sound-barrier breaker), flying in his North American Sabre F out of Hahn, Germany, had trouble handling in dogfights. It became a symbol of status for pilots of other countries if they could show film with a Canadian Sabre 5 or 6 in their sights.

Requests piled up for sonic booms at British air shows. When there were occasional complaints about such shattering noise eruptions, the Canadians chuckled that if the Brit air force were on the ball it would take credit for the mach-busting and claim they occurred during testing of the new fighter, the Hawker Hunter, that was long overdue joining RAF squadrons. The skies over England became a happy hunting ground. The only official exhortation was to avoid "Purple Routes," flights involving royalty. Hassles were the rule, young jet pilots—influenced by Second World War vets who had also flown Sabres against MiG 15s in Korean combat—earnestly throwing themselves into air exercises and mock combat, from high altitude down to the turf. Constant practice was the imperative for readiness.

Canadian Sabre pilots fighting RAF aircraft would never again encounter more competent and willing combatants. To tailchase an RAF Gloster Meteor, or de Havilland Vampires and Venoms with their tight turning radius meant the Canucks would almost certainly be wrung out like the washing. But the Sabres, with their vastly superior speed and diving stability, and three times the rate of roll of any RAF kite, were unbeatable in hit-and-run tactics. A Meteor could lose its tail by exceeding its mach limit; the Vampire could disintegrate.

The missions for Canadians flying out of Luff were myriad, but two things dominated their air time: the almost constant "rhubarbs" with British and a few American planes, and each other, and the almost constant battle with the elements. Sword jockies flew their day fighters with the mind-set of day-night-foul weather pilots. Instrument Flight Rules (IFR) approaches by Cathode Ray Direction Finder (CRDF) and Ground Controlled Approach (GCA) letdowns near and below limits were common.

Soon more than 1,000 Canadians had made themselves at home in a little county with a population of just 20,000. Many pilots affected the red-and-white scarves of nearby Stamford School because 410 Squadron's tail fin chevrons were the same color combination. As wives showed up, pilots and airmen secured new addresses: Lyndon, Melton Mowbray, Whissendine, Uppingham, Oakham. Personnel crowded Edith Weston's Wheatsheaf Inn, dubbed "Smokey Joe's," for ale, push ha'penny, and skittles. At that time, there was no Rutland Water lapping the Wheatsheaf's doorstep. The Horse and Panniers in North Luffenham village became The Nag and Bag.

Cold War fighter pilots, we arrived in North Luffenham in January, 1954, and two weeks with 410 Cougars was long enough to feel comfortable with the daily routine of operational flying, but never long enough to become insensitive to this England, its past and present. Some flights were unusually contemplative, fed by the aura of antiquity, the feeling that wherever we flew, the air still was eddying from historic events. We were the sharp end of an effort by the Canadian government to play a responsible role in the fledgling North Atlantic Treaty Organization (NATO). Canada's multi-billion-dollar undertaking was to build a significant NATO military force, including four wings of fighter planes in Europe. In England, Canada's 1 Air Division reported to 12 Group, RAF; on the continent, to Fourth Allied Tactical Air Force (4 ATAF), part of Allied Air Forces Central Europe (AAFCE). Supreme Headquarters Allied Powers Europe (SHAPE), the military command of NATO, controlled AAFCE.

The Sabre was the last jet fighter to depend on the pilot's eyeballs when engaged in combat. In future, electronic warfare, with its computers and long-range missiles meant combat could be initiated while the enemy still was beyond the horizon. Sabre pilots could recall the early flights by barnstormers and First World War flyers being described as "flying by the seat of their pants!" Well, today's fighter pilots say that's what Sabre pilots still were doing.

Though carrying what we considered pea-shooter type fifty-calibre machine guns, we had realized that the Hollywood-style dogfight, easy and tempting to fall into, was not an effective—or healthy—way to engage in combat. The margin for error was very small. The unheroic, ungallant hit-and-run tactic was the one we should employ, and did when we were seriously flying war exercises. But two considerations found us, more often than not, engaging non-Canadians in the old round-and-round, get-on-his-tail, tactics of the two world wars—the plain old thrill of it; and the need to hone our flying kills, to expose ourselves to the talents of non-tribe pilots and aircraft.

In heavy, laced-up jump boots, we clomped out to our aircraft in green G-suits, some with blue-grey cotton coveralls over the pressure-suits. We wore tight-fitting brown kid gloves (light protection in case of cockpit fire) and carried visored helmets. Each did his walk-around monitoring such things as movement of control surfaces; checking inside the nose air intake and the tailpipe; making sure the pitot tube cover was removed. Do up parachute straps over Mae West flotation vests (with a miniature hunting knife in sheath, to puncture the dinghy if it accidentally inflated in the cockpit). As we flew over the blinding white clouds, four pairs of eyes scanning in every direction for black spots, a tell-tale glint or the give-away contrail, joy welled, from the sense of accomplishment, the power of leadership. Sometimes, the emptiness of clouds struck one;

we were surrounded by it, but it was like being in a cotton bag. Our small shadows wisped on the far-off wall of the cloud, surrounded by concentric faded rainbows.

So, often as we could, we flew off in search of aircraft from other air forces that we could hassle with. "Tallyho!" was our exuberant call on sighting a bogey. (The air vice-marshal commanding RAF Twelve Group had told us that his heart swelled when he overheard Canadian pilots shouting "Tallyho!" He hadn't heard the call since the Battle of Britain). Pleased reverie followed landing. Every flight was metamorphic, the senses sharpened, focused and surging to a degree seldom known on the ground, your whole being opening up and, like an antenna, quivering to the other world, the world the non-flyer can never know. Any pilot with a soul felt it. Now, back on the ground, we stiff-legged away from the planes, once again feeling pity for the plodding groundlings who never experienced that other world above.

During Dividend, largest NATO air defense exercise ever mounted, No. 410 Squadron Cougars flew out of Coltishall, Norfolk, renowned Battle of Britain base for Douglas Bader and the primarily Canadian 242 Squadron. Front row: Wilf Thorne, Jim Shilliday, Steve Atherton, Don McCallum; Back row: Bob Morgan, Ev McKay, Ira Creelman, "Moose" McElmon (on author's shoulders), Bruce McLeod, Ray Oldfin, "Tweedie" Vaesen, Bill Johnson.

Later, we sat down at flights to study film, usually the result of a 50-minute operational practice flight, four aircraft in battle formation. The whir of the projector was an understated background sound, hardly furious enough for the action on the screen. The RAF, or USAF, or French, Belgium or Netherlands quarry banked and pulled left, vapour trails gushing momentarily from its wing tips; flipping to the right, gunsight switched on, diamonds of the reticle leaping into view on the windscreen in a perfect circle, the glowing pipper in the centre. Closing on the prey. Twisting the radar gunsight control, seeing the reticle expand and contract, searching for the target so it could lock on, more wingtip vapour trails erupting and presenting a grand plan-view against clouds whipping from left to right, a perfect target, the pipper centred on the canopy. Pull the trigger. One "dead" pilot; one "destroyed" Raff Meteor. A good feeling, even if it couldn't approach the exhilaration—or horror—of the real thing. We would leave the eight-millimetre reel on the projector for the corporal to file on the rack, satisfied our aim was improving, our ability to manoeuvre a gun platform into position was showing promise.

Constant practise: GCAs and targets of opportunity; cine gun exercises; low-level nav map reading; high-level formation and tail chases; aerobatics, practise forced landings; Battle of Britain flypast practise; sector exercises with exciting scrambles; low-level Rat and Terrier exercises with the ground observer corps providing information we used to chase invaders; demonstration ground attacks; air-to-air gunnery at RAF Acklington and over The Wash. We flew English Electric Canberra interceptions. On one of these, the Mk. 2's published service ceiling of 47,100 feet was verified. On the tail of a climbing Canberra (possibly from Scampton or Cottesmore), camera grinding away, my aircraft was beginning to mush at 47,000 feet, tailpipe temperature entering the red.

Another Canberra interception over The Wash was unintentional. Lollygagging along, I looked down and wondered at concentric circles on the water's surface. The light bulb clicked on and my eyes strained to look up. Almost directly above me was a Canberra—with its bomb bay doors open. The Canberra, activated in 1951, could carry an atomic bomb to Russia. It was picked up on licence by the U.S. as the B-57. And the RAF Canberras used Nene engines, the ones that powered the early Russian jet bombers and fighters after they were gifted to the Soviet Union. Warplanes on both sides of the Iron Curtain were using variants of the same engine.

One typical claggy day demonstrated the cool, no-nonsense attitude of RAF tower operators: pilots from all three squadrons were diverted to RAF West Raynham when Luff was socked in. There was a huge queue of RAF and RCAF planes circling. When a 439 pilot, short of fuel, finally was cleared to land, at first he couldn't get his wheels down and notified the tower. The tower controller, phlegmatically British and uncompromising, replied: "Canadian jet,

The rudder pedals…where are they? Dave "Crocodile" Alexander, in cockpit, and author Shilliday demonstrate the steep learning curve facing would-be Sabre pilots at Chatham, New Brunswick's Operational Training Unit.

we cannot accommodate wheels-up landings." What if the pilot had not been able to get his wheels down? Being Canadian, he probably would have landed anyway.

The One Wing Sabres at North Luffenham were held back from bases on the Continent, because they were part of Britain's first line of defence. The Russians by 1953 had A-bombs and their Tupolev Bear and Myasishchev Bison bombers could deliver them to Britain. The Canadians' job would be to stop as many of them as possible. The British—at the time—didn't have what it takes to fight a war. The penny-pinchers had almost flamed-out the RAF. In the meantime, Britain needed the Canadians. Her Vampires and Meteors were outdated. Britain eventually would buy some Canadian-made Mark Four Sabres, flying out of Linton on Ouse, in Yorkshire. Even the Sabres would be obsolete before long. But in those early days of the Cold War, they were as up-to-date as anything else in the air, a good match for the Russian MiG-15.

The RAF was heavily dependent on friends and had borrowed Boeing B-29 Superfortresses (which the Canberra would replace) from the Americans. The Yanks were flying Boeing B-47 Stratojets out of Sculthorpe, just northeast of Luffenham.

My Sabre coasted at 35,000 feet, eastbound. The smudge of King's Lynn in the Midlands drifted backwards to my right. Ahead would be RAF West Raynham and, a little north of that Sculthorpe, the American bomber base. Then a smile: a little farther east, RAF Little Snoring. A glint to port and the focus of my eyes ranged. One six-engine, B-47 Stratojet, ten low, our side's plane with the same job as the Russsian Beagle, only in the opposite direction. Mixed the controls, and my Sabre dropped and banked. With a 10,000-foot height advantage, it quickly overtook the bomber. Ahead, the huge warplane lost a little altitude and started conning, multi rows of churning white spewing back from its exhausts and creating an ephemeral highway in the sky. The Sabre dropped close to the condensation trails that whipped below at many hundreds of miles and hour. Descending to just above the cons, a burbling sensation followed by severe thumping turbulence while dipping through and clear of the jet wash. As always, the contrails were oddly disagreeable. The fact they were there, however short-lived. Earth was scarred with the marks of man that never could be erased: too bad he had to leave traces in the sky.

Closer to the wingtip of the bomber. This was trickier than expected. The big bird, precursor of the B-52, actually was flapping its wings. On the ground, the B-47's wings drooped with negative dihedral (anhedral). But airborne, at speed, they took on a positive angle, the wingtips moving up and down through an arc of at least eight or ten feet. I continued to fly straight and level, rather than attempting to stay with the bomber's wingtip. An eerie sensation, watching that whipping slab of aluminum. The bomber carried a two-man crew. It had a perspex cockpit similar to a fighter's, well ahead of the giant, swept-back wings from which hung all those engines. The pilot was unaware of the Sabre formating near his right wing. The second member of the crew sat behind, hidden under a canvas canopy reading instruments.

The bomber pilot's head swung left, then right, until his eyes caught my Sabre. Shocked, he twisted sideways to stare. The canvas cover popped back to give the other flyer a look. Fun over, I gave the thumb's up. Eased back on the stick and climbed away. It wouldn't be that easy against a Beagle. *But they're lucky I wasn't the enemy.*

Outwardly, of course, life in Britain was tranquil, nothing to compare with the total-war mobilization and tenseness of the Second World War. But, inwardly, the British and their NATO allies paid total attention to the possibility of military conflagration, and how to handle such a calamity.

DEFENDING AGAINST ATOMIC ATTACK

THE RAF'S NEWEST fighter, the Hawker Hunter, couldn't gather in great flocks because it still had the unprofessional habit of flaming out when its guns were fired, so Canadian Sabres were the spearhead defence against enemy atomic attackers (European, Canadian, British and American aircraft) invading England's airspace during a war exercise.

In Cold-War sombre July, of 1954, Dividend was the largest NATO air defence exercise ever mounted, with 3,000 aircraft simulating an atomic air attack on England. The prevailing military belief at Allied Air Forces Central Europe (COMAIRCENT) was that Russia had enough bombers in East Germany and other forward positions to attack most major allied targets in Europe, including those in Great Britain. What would we be defending against in a real war? Among the latest types of bombers in the Eastern Bloc was the Ilyushin-28, dubbed the Beagle by NATO, the equivalent of the British Canberra. Another was the Tupolev jet design, the Badger. Its civil version was the Tu-104 passenger airliner. On our side, from Britain, we could bomb with the Canberra and the RAF's strikingly beautiful delta-winged Vulcan. The Americans were flying six-engine B-47 Stratojets out of Sculthorpe, not far from North Luffenham.

Pilots of 410 (Cougar) Squadron were told that their role during the exercise would be to defend England's skies while flying out of RAF base Coltishall, in Norfolk, north of Norwich and just inland from the channel coast. Coltishall had been the Battle of Britain base for 12 Group's 242 Squadron, commanded by the legless squadron leader, Douglas Bader, and made up mostly of Canadian pilots in the RAF. So Cougar pilots' heads were really in the clouds at the thought of scrambling from such a storied location. We flew to Coltishall July 16. Spirits of *The Few* lingered everywhere on the runways and in the flight rooms until we returned to Luffenham on July 25. The Raff defenders did their best in two-engined Meteors (Meatboxes) and single-engine, twin-boomed Vampires. It was difficult not to feel superior. All three aircraft types had their recognizable idiosyncrasies—the Sabres, the black smoke trail that we joked about; the Meteors, a ghostly moan in the air from their Rolls Royce Derwent engines, heard before they were seen; the Vamps, a startling explosion out back of red-orange flame from their cartridge starters.

Mainly, the memory is of marginal flying conditions (the unfortunate ground crew lived in tents and the first few days rainwater was up to their ankles) and intense flying. Pilots often stayed in their cockpits through three or four sorties. Soon the practice was to cut engines and coast to the flight line where fuel bowsers waited. Nozzles went into the tanks when motion ceased, and riggers and fitters clambered all over.

We worked shifts. On July 23, I climbed into Sabre 321 and scrambled just after 4 a.m. No. 321 and I flew five sorties (an hour of that in cloud), the last one ending at about 11 a.m. During one of these trips, our leader, Pat Mepham, fell asleep in the cockpit while waiting for the next scramble. We other three trundled away without him and I'm sure he was mumbling something akin to the wartime RAF parody, "There they go, and I must hasten to catch them up, for I am their leader." He caught us up. That morning it was mostly Dutch and Belgian F-84 Thunderjets coming in. An incoming "bomber" could take mild evasive action but, once a defending fighter was on its "six" (tail), it had to fly straight and level. This allowed the defending fighters to move in close and take identification numbers, usually from the tailfin, which we proudly recited to the Intelligence Officer at debriefing. It seemed something more than make-believe for Cougar pilots when told we had "shot down" more attackers than any other defending squadron. We were The Chosen.

What was not make-believe during exercises was the conviction we all had that, when told to fly, we would fly, no matter what. We had heard all the stories of rotten weather and what sometimes happened to fighter pilots, how four RAF Meteor pilots had all been forced to eject when the visibility was zero at their base and they were short of fuel. Not long after arriving at North Luffenham, this acceptance of risk became more real. The second day into an exercise with RAF squadrons, Flying Officer Cal Drake and I were put on stand-by readiness. The Met briefing had been dismal, low clouds, fog and rain forecast for the rest of the day. The whole island was enveloped in weather below safe flying margins. So we were surprised, and somewhat dismayed, when the alert phone rang and we were sent out to our Sabres. The drill was to be ready to start engines the instant the radio conveyed the controller's orders to scramble.

We couldn't believe that they actually would scramble us in these unforgiving conditions. But we sat there, rain dumping over us, the end of the runway out of sight in mist, the clouds seemingly rubbing over our canopies, our ears two sizes larg-

er as we waited for the command to go. And it came. We started engines. We looked at each other, and Drake shrugged. But before we could begin taxiing towards whatever fate intended, the order to abort the flight was issued. We shut down again. Testing us? But we would have taken off. We would have decided on our subsequent moves when the time came.

Dividend was over. My beautiful Sabre leapt off the runway, and I banked to a reciprocal heading a little north of due west and held it at just 500 feet, a thousand feet beneath the broken cloud, forward visibility obfuscated at times by sunbeams squirting between the puffy formations above. The idea was to fly out over The Wash and then follow the pointing of the westernmost teat to North Luffenham. Trees, roads and streams ripped by below. Jet flying at its best, the most exhilarating—and the

Cougar "jockies" pose with a Sabre at North Luffenham: Front row: Bill Johnson, John England, Bruce McLeod (beneath intake), Fred Kozak, "Tweedie Vaesen", Steve Atherton; Back row: Bud Venus, Ira Creelman, Ev McKay, "Moose" McElmon (inside intake), Cal Drake (astride nose), Wilf Thorne, Pat Mepham, Al Tegart.

most unforgiving. Ahead, the rooftops of a large structure zooming in, pulled up a little, roared just above a multitude of chimney pots and, curious, began to circle for a better look. But then, the light bulb flashed on. Oh-oh! I wheeled back to the original course and scampered. Sandringham. The Queen's summer home in Norfolk, wonder if I woke up baby Anne! (Now, of course, The Princess Royal).

Near The Wash, my flight path was increasingly interdicted by bursts of rain and, finally, an impenetrable-looking black menace beneath towering storm clouds. I climbed to 40,000 feet in under four minutes. What a plane! The radio compass needle pointed at 255 degrees for North Luffenham. Occasionally it swung, distracted by the electrical disturbances in nearby cumulonimbus clouds, their wispy tops streaming out in the familiar anvil pattern. Dangerous for unwary pilots!

Below was a Grand Canyon, a Hell's Gate of such storm clouds now, some spaced three or four miles apart, between some of them a drop through empty space of 30,000 feet. A challenge. Stick forward, throttle to full power, dropped 10,000 feet, speed building close to Mach One, before pulling back on the stick and beginning the return climb directly towards the tallest cloud, dirty black at bottom, graduating to snowy blinding innocence at top. The Sabre bulled up the face of the formation towards that gleaming top, determined to best it and bunt over the peak. Up, up. The airspeed plummeted. Nearing the top, about 2,000 feet to go, I was rocking my body, trying to assist the Sabre over the top. The controls mushed. The mighty Mark Five wavered, wallowed. Just before it stalled, I eased the stick forward and fluttered mothlike, spent, into the enveloping white.

No contest! Never-apologizing nature could beat you any time. A humbling experience. After a procedure let down, back on terra firma.

GLIDE PATH

ROOKS SULKED ON high branches like black notes of a dirge, their fluffed feathers giving off no highlights on this grumpy, gray English Saturday.

"What's the weather?"

"Good enough. Ceiling's two thousand, viz three miles, a couple in occasional drizzle. There's worse coming, but it's still hours away. Let's go!"

That's a reconstruction from when we 410 Squadron pilots were flying out of RAF Coltishall, Norfolk, for Exercise Dividend, a giant simulated atomic attack from the east, in 1954. This was a non-exercise flight. "Moose" McElmon would lead. Time has erased who Three and Four were. We agreed to fly over to the seaside resort at Great Yarmouth, a short hop east. Then we could climb up and go hawking for targets of opportunity, probably scarce on a Saturday. Coltishall's wartime main runway pointed towards the channel, better to get the Spits and Hurries into the fray quickly. By the time our four Sabres had formed into loose battle formation, we were approaching the English channel coast across the expanse of Breydon Water at the edge of the Broads, nudging Great Yarmouth, where the hero of Trafalgar Bay, Lord Nelson, used to hang out with Lady Hamilton.

The cloud base was lower than expected at 1,500 feet and we stayed clearhood, just under it, thundering over the resort that had been so badly pounded by German bombers a few years earlier. We plunged out over the white-capped seas south of the town and immediately McElmon had to take us down to 1,000 feet to stay under the clag. He circled to starboard, descending, until we were just off the mainland, heading north towards Great Yarmouth again and at 200 feet above the crashing waves. Number Four was lowest man in the formation, and closest to the menacing turmoil below.

Military jet flying was un-time travel: no tomorrow, no yesterday, just now, right now. You must do it right. Now. We pressed on, McElmon gradually settling his plane lower, just above the sea, so close, in fact, that moisture was sucked into my intake and issued in a fine spray from the heat-and-vent system. Something dark flashed by, no doubt a seagull. That could have been it. They can leave a great messy hole. We all knew this was a game of chicken (gull?) and said nothing. And when we saw four aircraft almost directly in front and heading our way, we didn't have time to say anything. The RAF Meteors, possibly playing the same game, but unaware of us, moaned just above our four Sabres and were gone. If you don't know what's going on, everything is okay.

Ahead, dark writhing wisps, rainfall trying to stitch the clouds and the sea together.

"Red Section, line astern, line astern go." As he spoke, McElmon began a turn to starboard looking for more friendly airspace, openness. But weather was closing in fast. "Red Three, you and Four climb to altitude and check the weather at base. Call on top. We may have to GCA when we get back. Red One and Two will be behind you. Close it up Red Two." I tucked in as we started a climb through increasingly dark cloud and turbulent air.

"Red Three and Four on top." McElmon acknowledged. Then we burst out of the cloud at 35,000 feet, momentarily blinded by the sun and its reflection from the cloud tops. We continued flying south for another couple of minutes to provide clearance for Three and Four, then swung around onto a heading for Coltishall. The cloud tops were rushing just below us, occasional pillars of white nothingness getting in the way but offering no resistance.

"Red Two. You'll have to be boss," McElmon suddenly called. "I've lost my pressure instruments. Guess the pitot tube is frozen. (The pitot tube, sticking out of the leading edge of the right wingtip, admitted both dynamic and static air pressure for instruments, to show such measurements as altitude and airspeed). I'll hug you all the way down, baby."

"Rodge, I'm lead." When McElmon was on my starboard wingtip, I held up four fingers, then switched to Channel Four, the ground control frequency. He checked in. At the beacon, the pair of us started down. At 15,000 we turned onto a reciprocal heading, toward the runway. Ground Controlled Intercept had us. GCI positioned our aircraft on the plan position indicator scope, reported our position, heading and altitude to the Ground Controlled Approach operator and handed us off.

Flying GCAs had great appeal, your appreciation at its highest when they were successfully completed. This one was dicey. The operator had reported half a mile or less visibility in light rain and fog. Cloud ceiling was 200 feet. Red Three and Four had made it down in slightly better conditions. At this rate, the runway could be completely socked in by the time we got there. The descending formation was now electronically clear to the GCA controller, but to me the only clarity was the cramped world of my cockpit. There was a grey blanket over the windscreen, little light. My eyes were moving over the instruments continually, messages going to my brain. My hands mixed the controls: on and off with the throttle; this way and that on the stick, responding at once to the controller's words and the instruments' information. Nothing focused attention better than the mental image of wreckage and flames at the end of the runway. Now the controller had us on the glide path.

"You are fifty feet below the glide path. Three miles to go. Decrease rate of descent a little…two-and-a-half miles to go. Resume normal rate of descent. You are now on the glide path. Check wheels down and locked." All the patter came from the controller. Following procedure, he had instructed me to say nothing further. Before long, he would tell me to take over visually and land, and at that point I hoped the ground would be visible. If not, we might have to overshoot, climb up out of this mess and pull the handle.

"One and a half miles. Slightly below the glide path. Reduce rate of descent slightly…on the glide path. One mile to go. Lined up with the runway…half a mile. On the glide path. Two thousand feet. On the glide path. One thousand feet." A pregnant pause, then, "Take over visually and land straight ahead." We still were in cloud. No we weren't. We were on the runway—well 30 feet above it. We were beyond the button. The canopy still was bumping cloud. I eased the stick forward pulled off the throttle and gently braked. McElmon was beside me, his masked face staring at me as though holding onto a lifeline.

Instrument flying—nothing like it! Good for the health.

FIRST SABRE NIGHT FLIGHT A CLOSE THING

A TAPERED BRICK SMOKESTACK towering 300 feet broke the horizon at Ketton, about three miles off the east end of Runway 26, and was affectionately referred to by returning pilots as the 'Luffenham Beacon.' Pilots could see the chimney tip poking out nonchalantly when the ground was hidden by a layer of low-lying cloud. We had compass and radio aids, sure, but the pipe was a friendly, reassuring sight.

Unlike some of their wartime predecessors, our young flyers were getting the time to sharpen their skills. Their official status was "operational" and they were becoming exceptionally good at their job. In the Cold War RCAF, before the all-weather CF-100 Canucks debuted, Sabre jet "jockies" were expected to fly their day fighters at any time, unintimidated by weather or dark of night. They were scrambled on exercises in rain and low cloud ceiling, sometimes having to divert to some distant Royal Air Force field because they could no longer land back home. They came down from the stars in battle formation, landing in darkness like night fighters.

I had been with 410 Squadron two weeks, fresh from Operational Training Unit at Chatham, New Brunswick. Early Monday evening, Jan.18, 1954, I was being briefed for my fourth flight since arriving by Flight Lieutenant Fern Villeneuve, of 441 (Silver Fox) Squadron, practise interceptions during an official exercise. Before the day was out I had teetered on the edge of that chasm that can open with such stunning abruptness in front of pilots when they extend themselves, so often in conflict with the immutable laws of nature. Nature always sides with the hidden flaw.

Villeneuve, who would become the first leader of the famous 1960s Golden Hawks Sabre aerobatic team back in Canada, and later have his image on a Canadian theme coin, said we would contact radar control after getting to altitude and do practise interceptions, which amounted to exercise for the radar controller and a test of the eyesight and alertness of the pilot being vectored onto the "enemy." It was 4 or 5 p.m. when I walked out to Number 159. Pre-takeoff checkon the mnemonic HHHTPFFGSO: hydraulics, harness, hood, trim, pitot, fuel, flaps, gyros, switches, oxygen.

Leader glanced across, made a circular motion indicating advance throttle to full take-off power. Down the runway, noses up, lift-off, wheels up, flaps up, into the climb flying as one, dirty JP-4 exhaust blasting from our tailpipes, filth, evident only in the Mark Two, that left a black-grey signature line for miles across the sky that, pilots joked, cost three dollars a yard. The two jets gradually faded into the blushing west, diminished not only by distance, but also by gathering clouds piling ever higher and promising a premature twilight.

Working with the radar controller, we completed several pilot interceptions on each other, the last proving difficult because the light was fading. I located the switches and turned on the cockpit lights, suddenly realizing that this was my first Sword flight at night. Still a half light at altitude as we headed back to base on radio compass, enough light to outline the sulking cumulonimbus clouds jostling a few thousand feet below, dark ragged pillars seemingly waiting to engulf our puny planes and play catch with them. We started a procedure let-down, 350 knots indicated airspeed, leader on the dials and my eyeballs, pupils dilated, fastened onto his aircraft, now not much more than a silhouette in the thickening murk.

The sudden appearance of a ghostly envelope engulfing leader's aircraft awed me, a sharply undulating white mist ballooning and contracting over the surface of the lead Sabre— shock waves were piling up, outlined by condensation. We switched to tower frequency for weather. Ceiling 1,000 feet, rain, visibility two miles (Hah!). As we approached Luff I was sent line-astern. A vast loneliness engulfed me. We had descended into dead night. The red tail light of the Sabre ahead was obscured by rivulets of rain twisting stickily down and across my windscreen, splayed and squashed by the tremendous pressure of the slipstream. I could see now, dimly, the lights of the station and runway approaching ahead and below.

At the break, the lead aircraft entered a steep bank to the right through 180 degrees onto the downwind leg. I followed, dumping undercarriage and 15 degrees of flap. I could see ahead the red light on his tail; below to the right was the line of runway lights. But another glance ahead and the red tail light was gone, swallowed in the clag. I'm no longer formating and my eyes dart to the instrument panel. Up to me now, have to orient to the dials.

Tiny dim-red bulbs illuminated the instruments. They showed clearly which way the numbers and needles were moving. Holy…! No altimeter. How high am I? The altimeter was a black hole, the bulb burnt out. A struggle to mentally adjust. I glanced down at the runway lights in the sodden mess below. What the hell! The line of lights was shortening; only two or three…then there were none.

I sat, engulfed in black, the cockpit interior faintly orange-pink—except for that black hole where the altimeter should be telling me how high above the ground I was. I decided against calling the tower; it was up to me. The plane ahead would bank to the right in a descending turn onto final. Then I heard my leader notify the tower on final, three greens down and locked. Prudent thing would be to continue my downwind leg until there was no danger of turning into the lead aircraft.

Alarm squeezed the throat but was deep-breathed away. Holding direction, maintaining airspeed, but how high? Tried to gauge the altitude by checking changes in airspeed and by watching the vertical speed indicator and the artificial horizon. But the VSI was hard to trust. It was a static pressure instrument subject to lag, and when you started a descent, it registered; but you could move into a climb and, for a few seconds, the VSI needle still registered a descent.

Time. I eased the Sabre around to the right, gently descending (I figured) through a half-circle turn until bound again on a heading of 260 degrees. I couldn't see anything

Author enjoying well-deserved break.

Night flight illustration: Dave O'Malley/Vintage Wings

ahead or below. I flew on, nervously checking instruments and the darkness ahead. My concentration was so intense, I could only mutter, "Final!" to the tower.

A light! A red light! It sifted casually from right to left and was gone. A red flare? The tower telling me to overshoot? Then I froze—that light had not hung enough, did not have the trajectory for a red flare. It had just appeared and gone, like a reflection on the canopy. That's what it was, a reflection of the red light on the Ketton smokestack about three miles off the east end of the runway. I had just flown past the brick structure, missing it by how many inches? Close enough for the canopy perspex to pick up the reflection and throw it at me while passing by—a casual wave from Death. That light was 200 or 300 feet above ground.

Abruptly, my senses registered new alarm. How can the black below get blacker? It seemed to be vibrating. Even as horror gripped me my left hand already had slammed the throttle full forward. The engine yowled and as the Sabre labored sluggishly, I picked up the undercarriage, the flaps. The blacker black, the vibrating black, was the ground, the treetops. The plane had been settling right into oblivion. I could sense the menace of a vengeful presence reaching, testing, expectant—eager to rip and devour this insignificant being so smugly intent on being its master.

I flew straight on, knowing the Sabre must be gradually climbing because the VSI was up all the time and my speed wasn't building much. I eased the throttle back a little, straining to see some approach lights ahead. I was close to complete disorientation when...there they were, those beautiful orange approach lights, advancing out of the rain and mist. Back with the throttle, wheels down, a little flap. Now, there's the runway, dimly illuminated by one row of lights along the left side. The plane was too high. I chopped the throttle. Really hot—still at 185 knots, about 60 knots too much. The runway was sliding away beneath me and I shoved the aircraft down heavily and braked, released, braked, released, braked...but the heat had built up and, as I turned off the runway and taxied towards the Cougars' apron, the tower told me casually that my main tires were burning. Didn't really care that much. I was back.

I parked and the handlers snuffed the fires. They were not impressed.

The Engineering Officer told me that, because North Luffenham was a wartime bomber airfield, its runway lights had been constructed to not shine more than necessary as a landmark for enemy night fighters and bombers. They had never been modernized. The lights were capped on top. They were covered on the sides for about 280 degrees, meaning that less than 90 degrees gave out light. The lights on one side shone towards one end of the runway, the other side shone towards the opposite end of the runway. On the break, things would have been better if we had turned left rather than right.

Fifteen minutes night flying is credited in my logbook to that long-ago flight. The most concentrated quarter of an hour of my life, I'm sure. The closest thing yet to the "blind flying" heroes of my boyhood. And too close to being remembered by the locals as "The bloke wot 'it the chimbley."

 A section of 410 Squadron Sabres at No. 1 Wing, North Luffeham, England, lifts off on scramble, the undercarriage already cycling.

long-ago flight. The most concentrated quarter of an hour of my life, I'm sure. The closest thing yet to the "blind flying" heroes of my boyhood. And too close to being remembered by the locals as "The bloke wot 'it the chimbley."

ALL AROUND, ETERNITY'S HORIZON

LIFTING OFF, SEVERING THE link with Earthly things was energizing. As ground clutter eased its clutch and fell away and the horizon tidied up on a curve, the pulsing of the jet engine synchronized with the heartbeat, a feeling of release. The flyer—bee, bird or human—learns the world holds more than might be expected.

Flying, the ultimate in escapism – this kind of flying anyway, pure and easy and smooth and relatively quiet, free of the shaking and mindless howling of the reciprocating engine. The sensation of lifting off less intense now, as the climb continued, as the mass beneath faded, the scattered clouds came down then sank below. As Earth's features became indistinct, it diminished in importance, its imperative pull overcome. All around, eternity's horizon.

An air test. The Sabre had been given a new engine and needed checking out, normally done close to the airfield, but why not visit a German tailor on base at No.3 Wing, Zweibrucken. Sign out for the test and a navigation trip, filing a flight plan NL-XP-NL—North Luffenham, Zweibrucken in West Germany, and back again. Climbing southwards on the way to Germany, the early morning sun beaming from the left, the low angle of the rays gave a different perspective to the scene below. To port, about 80 miles from base, the faint runways of Stansted Mountfitchet where the RAF sent its bent Sabres. At 42,000 feet and levelled out, a crusty grey splotch—London—passed beneath the starboard wingtip. Bank left a few degrees and press on, soon making out the thin white line of the Dover cliffs and the English Channel, even from this height bluish and ruffled by wind and current with coruscant whitecaps randomly visible here and there.

The engine sounded good, prompting a smile, because so many new pilots, those with faint hearts, reported engines running rough immediately they were over deep water. Mine pulsed on and all the needles were in the right places. The coastline of the Continent passed below and then, through a haze that hugged the ground, another crusty grey area amongst the muted browns and light greens. Pivot on a point above the Eiffel Tower and head for Zweibrucken. No.3 Wing, Zweibrucken, sat on a hill, just north of the French border, in a region that once belonged to Sweden, between the RCAF bases at Grostenquin in France, and Baden Soellingen, West Germany. Fifty minutes after take-off, on the pitch at Zweibrucken, the nest of 413 Tusker, 427 Lion, 434 Bluenose, all Sabre squadrons, and later 440 Bat (CF-100) Squadron. While the aircraft's fuel was being topped up, walked to the tailor's shop, got measured for two suits, ready in two weeks. Back to the waiting Sabre musing that the flying had to be done—just took advantage of an opportunity.

Over home base was a good time for practice, a simulated emergency landing. Switching to tower frequency to inform the air controller, I selected the new channel by touch, rather than looking down to my right side to see the channel switch. Crashes had been blamed on pilots becoming disoriented while flying in cloud, after taking their eyes from the instrument panel to switch channels. Now for a spiral down over the field and maybe a "touch-and-go," an overshoot, before landing. But the tower cautioned that fog smothered North Luffenham and visibility was below limits. It would be necessary to divert to RAF Wittering, about 10 minutes east. Wittering dated back to 1916, when Sopwith Camels and Pups droned from its grass runway; by 1939, Spitfires were operating there and engaged in the Battle of Britain. Canberra jet bombers began flying out of Wittering in 1953, and Valiant V-bombers in 1954. Later, the marvellous Harrier jet fighter was stationed there.

After landing on Wittering's new concrete runway, the Sabre (an object of great interest to the RAF erks) was topped up. When the tower said weather was okay at North Luffenham, we were off for home again. No answer after calling Luff tower for pigeons (a compass bearing). After several tries, the earphones crackled: "Aircraft calling North Luffenham tower, if you read click your transmitter once." This was done. "Roger, aircraft, from now on click once for yes, twice for no." I pressed once. "Roger. Are you experiencing any difficulties?" Two clicks. The operator then gave a direction to steer for base, and soon the Sabre was parked in front of 410 Squadron's dispersal. The alertness and smarts of Luff's tower operators always impressed.

MID-AIR COLLISION

WHEN YOU ARE A PILOT practising for war, much of the time you fly to fine limits. But the line between skilled airmanship and reckless bravado can be extremely thin. For Sabre No. 180 and me, on August 17, 1954, it was just millimetres.

Wilf Thorne and I lifted off as Blue Section from North Luffenham's Runway 26 that afternoon eager for targets of opportunity (no hated directives yet against dog-fighting from AFHQ) and then a procedure letdown and Ground Controlled Approach. We hassled (fought) some short-nosed RAF Meteors (they turned so tightly), bounced a couple of our own Sabres, and finally headed home. We began a procedure let-down in close formation. My eyes were in the cockpit scanning instruments; Thorne's were focused on that point of my near wingtip where the pitot tube met the leading edge. By positioning his aircraft to do

⟫
"Bounced" by a Sabre pilot whose ambition was greater than his skill, the author's aircraft sustained scrapes and dents that—for the sake of a few millimeters—could have been total disintegration.

that, his formation flying was impeccable. We descended towards the top of the flat cloud table at 25,000 feet, just a couple of thousand feet below, focused but relaxed, lulled by the satisfaction of a productive sortie and pleasant flying conditions. Airspeed, altimeter, artificial horizon, compass, a circuit scan of the instruments….

"WHUMP-BAM!"

Between blinks, discombobulation. I looked up, still feeling the sharp impact. Directly in front was the plan-view of a Sabre. A huge bird, from left peripheral-vision to right. Just as suddenly, it was diminishing, rising and pulling away. Heart pounding, my eyes scanned the instruments again, looking for problems, danger signs. We still were descending. Thorne clung to my wingtip. I pressed the transmitter button: "Get his number?"

"Negative."

"I was hit. Take the lead. Check my aircraft." Thorne's aircraft dropped down, crossed under and rose on my left. "Looks okay." Cloud deck coming up, then we were in. I settled down, formating on Thorne's plane, heartbeat slowing.

"M-m-m-ph!" The oxygen mask suddenly sucked tight to my cheeks, and I was gasping. With my left hand, I ripped the mask loose, inhaled cockpit air. We still were more-or-less formating, but I decided to call Thorne and tell him I was breaking off. Sliding out of formation, with my left hand I raised to my face the oxygen mask that contained the microphone: "Blue Two breaking off." The loose oxygen mask translated my message, something like: "Boom doo enky aw." Breaking free of the formation while still in cloud, I had to re-orient myself on instruments quickly. Continued the descent, completed a procedure turn onto a reciprocal heading and broke out at 3,000 feet.

Thorne had come out somewhere ahead. He called clearhood and homebound. He couldn't make out my reply, that seemingly was interspersed with the impolite sounds of flatulence. "Romboo (phrrrtt) aw-phump (flbflbflb) fipoo." If Thorne was concerned, at least Shilliday was still making sounds. As I taxied in and parked, the engineering officer and several "stableboys" were checking and pointing at various parts of the aircraft. Scrape marks traced the path of the collision, starting on the port side of the fuselage, just ahead of the wing root. A heavy indentation extended ahead to the oxygen access cover. This was buckled. A chunk of foreign material, gray and flaky, stuck out. The scrape continued to the nose air intake, which was slightly bashed in. The engineering officer shook his head: "You're a lucky sod."

Driving the squadron pickup to the huge brick hangar at the other end of the runway from ours, I parked and strolled down the flight line, examining the parked Sabres. The last one was the culprit. The tail fin normally was topped by flush fiberglass that covered an antenna. This whole area was missing and explained the piece of extraneous material sticking from 180's oxygen access cover. A mid-air transfer.

The aircraft had been flown by Flying Officer F. I wrote an accident

| 128 | A MEMORY OF SKY

report, as did Thorne. F, a Hotspur of the Heights, was banished to Air Div at Metz to fly a desk for six months, following an inquiry into the "C" category (fixable) crash. A year later, Sabre 180 was sold to Turkey and was in an "A" category crash—flying out of Eskisehir the pilot ejected at 9,000 feet after his controls froze.

AIR SHOW

WE WERE TREATED TO EYE-POPPING aviation displays that September at the Farnborough Air Show, with cutting-edge military aircraft and flying displays for hour after hour, as a country's latest aeronautical accomplishments were exposed to critical eyes from many nations. The giant Vulcan V-bomber was flown to its limits. The Hawker Hunter showed off, and it wasn't many months before we were butting heads in the sky. We were impressed by the prototype Canberra bomber that had just made a record-breaking flight across the Atlantic, and the Blackburn Beverley freight transport that could load huge military trucks.

The tiny Folland Gnat prototype made a pitch for success and fame that never panned out. Possibly copying the Gnat idea, the U.S. came out with the diminutive Northrop F-5 Freedom Fighter in an attempt to challenge the "size-weight-complexity-cost spiral" in single-seat fighter design. Now well entrenched in dependence on U.S. aircraft designs, Canada arranged for Canadair to produce on licence 204 CF-5 Freedom Fighters from 1968 (another Hellyer brainwave) to 1975. But their short endurance, unfitted for Canada's vast distances, resulted in the air force eventually mothballing most of them. The plan had been that they would be available for fast dispersal to NATO service in Europe, mainly Norway, through refuelling by Boeing 707 aircraft, but ended up simulating enemy aircraft in war exercises and as a trainer in tactics before pilots went on to the CF-18.

The 1955 Farnborough Air Show featured a Canadian Avro CF-100 Mark 4B Canuck, powered by two Orenda 11s, the first military jet designed and built outside England to appear at the show. Starting in 1950, 692 CF-100s were built, as many as 13 squadrons were equipped with them, and they were flown in the Canadian air force until 1981.

The last Canadians left North Luffenham for the Continent in 1955. From 1959 through 1963, Thor missiles were placed around the area.

Counting the hits after air-to-air firing at Rabat-Sale, Morocco. Point Five ammunition was dipped in coloured paint that left a discernible mark on the nylon mesh, and pilots checked their colours carefully to obtain a score and establish who was "top gun".

JIM SHILLIDAY | 129

In late 1997 the base at North Luffenham, still remembered with pleasure, and remarked on with pride by hundreds of ageing Canadian pilots and ground crew, was officially closed.

BADEN SOELLINGEN— SHOOTING (AND DODGING) THE FLAG

AFTER 10 MONTHS AT North Luffenham, 410 Squadron left "Blighty" and spent six months on temporary duty at Four Wing, in West Germany, waiting for our Marville, France, airfield to be completed.

We finished milk shakes at the 4 (Fighter) Wing snack bar at Baden Soellingen, West Germany, went back to flights for parachutes and crash helmets, and strode to the oak-tree-surrounded dispersals where our Sabres loitered. Soon after 10 a.m. on February 2, 1956, four sections of 410 squadron aircraft were airborne and pushing south for a three-week gun-shooting spree in North Africa.

Sixteen planes took off from Four Wing, leaving behind the showpiece of Canadian bases in Europe. We had come to Soellingen in the fall because our base at Marville, France, still wasn't ready. Soellingen is in southwest Germany, in Baden-Wurttemburg, the third largest state in the country. The wooded hills and picturesque villages charmed us. Baden-Baden was the main town and home to centuries-old spas, posh hotels and the Spielbank gambling casino, frequented by potentates, sheiks and general high-and-mighties through the decades. The nearby air force base was permanent home to 414 Black Knight, 422 Tomahawk, and 444 Cobra, all Sabre squadrons, and later, 419 Moose (CF-100) Squadron.

Our jets rolled down the runway in pairs until all were airborne. They joined up in a loose formation of four-plane sections and climbed southwards, some of the pilots tuning in Radio Luxembourg for relaxing popular music. We overflew Geneva, Grenoble, Avignon, crossed out over the Mediterranean and circled back above Istres, the French air force's experimental-flying base (where they had a varied collection of captured Luftwaffe aircraft with which they still were experimenting) near Marseille, switching to tower frequency, 118.3 mc. Looking down, I was fascinated but somehow apprehensive: above the Gulf of Lyons was an unusual cloud formation, a roiling pinwheel of expanding circles covering many square miles, graphic indication of a low-pressure area.

Our leader called Istres Tower, his voice measured, calm, "Overhead with sixteen. Landing instructions, over." A Canadian pilot had been temporarily stationed at Istres to assist the French controllers, even though English was the language of international aviation. Fluency was a problem. The tower replied. Another calm voice, but heavily accented. "Istres runway closed. We have the mistral. Blowing up Rhone Valley at sixty nautical miles, gusting eighty-five." He paused. "Go civil airport Nice. Over."

When Nice reported that it, too, was closed, our leader ordered his pilots to update their fuel situation so he'd know who had the least fuel left. "Istres, Blue Lead," his voice less measured now, "where's the Canadian controller? Find him. We're coming in. We're at forty thousand. Eight minutes fuel." There was little margin of safety. Each pilot was aware he had to get down and land successfully first time. None was likely to have enough fuel for an overshoot. In such weather, and at a strange airfield, anxiety could build.

The four-plane sections began high-speed spiral let-downs at 20-second intervals, popping speed brakes, retarding throttles and pointing the aircraft steeply downwards. By the time they were approaching the runway, which was laid down on a bearing of 338 degrees, the Canadian controller was on the radio. Luckily, the mistral was blowing straight up the runway, not across, after its violent tree-flattening journey across Provence generated by the venturi effect of the narrow gap between the Pyrenees and the western Alps. Against this fierce wind, the aircraft seemed to hang above the button when they rounded out. On touchdown, power had to be applied immediately to stay ahead of the next aircraft landing on the 11,000 foot strip. After leaving the runway, the Sabres taxied a mile back to the tarmac, bucking a severe cross wind. This required constant heavy left leg pressure on the rudder pedal that activated the nose-wheel steering. When climbing down from his Sabre, a pilot's left leg tended to give way.

Author landing Sabre on the button of Sale's French air force runway, after air-to-air gunnery over the Atlantic. Air France DC-4 prop-driven airliner waits for taxi clearance.

The French pilot and writer (*The Little Prince; Night Flight*), Antoine de Saint-Exupery, had been stationed here during training for his final air force flight tests. He called it *bagne des eleves morts*—penitentiary for doomed greenhorns. He also had flown out of Rabat-Sale, the French air force station in French Morocco, where we were headed for another session of air-to-air gunnery practise. A couple of hours were spent refuelling men and machines at Istres. The entire flight, from the Swartzwald (Black Forest) to North Africa, was flight-planned for 1,200 nautical miles.

The flight to Africa was routine, except for Flying Officer Ev McKay's departure from the formation with some engine problem. He descended to land for repairs at Oran, on Tunisia's coast (where Churchill ordered the destruction of the French fleet, in 1940). We flew west of Majorca (tuning in a Valencia radio station for a fix), skirted the eastern shore of Spain (the Pyrenees, Spain and Africa all were visible) then flew southwest until the Atlas Mountains of North Africa appeared. Only dust and rock flowed in the heat waves of the many dry hill rivers. But farther west, as the ground sank towards the Atlantic Ocean, green appeared and the Sabres passed through small rain clouds as they approached Sale airfield. The squadron landed at Rabat-Sale mid afternoon, three flying hours after take-off in Germany. Rolling back our canopies, the blasts of hot wind seemed exotic.

Four 441 Squadron Sabres show how they would fly echelon starboard on their way to air-to-air firing over the Atlantic range of French Morocco.

The Canadian air force ensign, a red maple leaf centering the roundel, a Union Jack in the top inner corner, flew in front of a low brick building, the pilot's flight room. Greeting us was wagging, gamboling 'Sabre,' a particularly large Great Dane owned by the unit commander, Wing Commander Bob Davis. Also welcoming us, but not so effusively, was a large, free-ranging turtle with the number of the previous visiting squadron painted on its carapace. We wandered in, stiff and curious, stashed equipment in lockers and helped ourselves to soft drinks and chocolate bars supplied for an energy lift. Then we were driven in a small bus to a one-storey barrack of stucco, painted in the yellowish-white of equatorial countries. Inside, the accommodation was clean but Spartan, two-bed rooms whose windows overlooked shading balsa trees. Light meals we had at the French officers' mess, but we ate dinner at the Air France terminal, our first one a good meal of African lamb cous-cous, with a "verrry" hot sauce, said the grizzled waiter who was dressed in white duck draped with gold rope, like an admiral, and told us the meat was camel.

First time towing the flag for air-to-air gunnery was something like flying your first solo. Pushing the envelope. You taxied onto the end of the runway to a target flag lying on the concrete. Six feet wide and 30 feet long, its front end was secured to an iron bar to provide stability, to make it "fly." The nylon weave was threaded with aluminum strips to act as a reflector for the radar gunsight. The flag was attached to 100 feet of cable. The ground crew attached this cable to the tow aircraft after the dive brakes on the fuselage, near the tail, were opened. The cable was hooked in the right side and the brakes were closed, securing the line. It was considered wise not to take-off under 180 knots before lifting off: hot African air was thin and the Sabre could stall at a higher airspeed than usual. The climb was steep to ensure the flag behind didn't catch on anything.

Three aircraft were fuelled and armed. When the tow Sabre began to roll, the flight leader gave the signal to start engines. They would take off several minutes later. The tow aircraft's speed would be restricted because of the flag, and needed time to position itself over the Atlantic Ocean firing range before they joined up. Airborne, the three shooters headed north towards Port Lyautey on the coast, turned west and flew out over the ocean range, soon catching up to the tow plane.

"Blue Tow, Blue Section five back."

"Rodge, Blue Lead. Have you. Arm your guns. Report on the perch and I'll tell you when to start."

The "perch" was a position even and parallel with the tow aircraft, 1,000 feet above and half a mile or more to starboard. The three shooters were line astern. When the first aircraft left the perch to begin its diving attack, the next aircraft would move up to the vacated position.

"Blue Section, Blue Tow. Clear to begin."

"Blue Lead in." The lead aircraft banked sharply, began its dive and reversed its turn so that it was executing a curving attack from the rear starboard quarter of the tow plane. The second plane throttled ahead to its place on the perch.

"Blue Lead off." The first plane reversed its direction after its firing run, began a 360-degree turn and climb back up to position.

Everything went so fast. Pilots had to take care not to fly into the flag. But they wanted to do this right, to get a good score, to win the other pilots' respect. As they dove from the perch, the flag was far away. And so small. As it grew slowly in size, the pilot operated the radar gun sight, watching the lighted circle of the reticle on his windscreen expand and contract as it searched, then locked on to the flag. Now the flag grew in size rapidly. The pilot could see it undulating. The flag was shortening as his path of attack more closely followed its flight path. And then suddenly it was expanding, jumping towards him. He could see the tail end of the flag waving violently, daring him to get closer. He yanked the stick back and passed just behind and above the malevolent target.

"Blue Three in." The attacks continued, several more seaward, then the turn around and only a few more landward so there would be no chance of firing too close to shore. The chunk, chunk, chunk of the breach blocks told the pilot his guns were empty. The belts of point-five ammunition had been dipped in coloured wax to coat the tip of each round. This left a coloured trace on the flag as the slug passed through, to identify the shooter.

The tow plane approached the airfield, flag streaming behind with all those little coloured marks that would make the shooters either proud or embarrassed (when a shooter's bullets hit the cable and parted the flag from the tow plane; the other pilots were outraged—they were sure that they had "really clobbered" the flag that time). Now the trick was to make sure the tow plane was far enough inside the airfield perimeter fence before dropping the flag. Sometimes, it was released too early and fell outside the fence. A Moroccan invariably was waiting to gather it up, put it on his burro and then demand 10,000 Moroccan francs, about $30, for returning it. Then the "gun plumbers" collected the flag and counted the scores.

In 1955, F/O Jerry Westphal, of 441 Squadron, scored 94 percent air-to-air with his 50-calibre Colt-Browning machine guns, a record that may never have been broken. How good were Canadian Sabre pilots at gunnery? Let Larry Milberry tell it in his *The Canadian Sabre*:

"When you come right down to it, the best fighter pilot is the one who can shoot best." In Europe during the fifties, with every country, "good guys and bad guys," fairly bristling with fighter squadrons, the very best were the Canadians. Year after year in NATO they walked away with whatever gunnery trophies were going. In particular, they seemed to have a monopoly on the coveted Guynemer Trophy, named for France's great

First World War fighter ace, Georges Guynemer.

"For several years, staring in 1958, NATO's best fighter pilots would meet at a French base, Cazaux, near Bordeaux, for grueling shoots to determine the Guynemer winner." There were British Hunters, American Super Sabres, Belgian Thunderstreaks, French Mysteres and, "to the dismay of all, The Canadair F-86s flown by the pilots of Air Division, ready, willing and consistently able to "wax the fannies" of all challengers. It got so maddening for other NATO teams that one NATO air force spent an entire year specially training a team to compete at Cazaux, but still was whipped by the Canadians."

ANTICIPATING 9-11

ATOP THE ANCIENT HASSAN tower in Rabat, our eyes wandered: above the guide's head, high in the sky, silhouetted against a deep unsullied blue, the lazy stork a moving splash of white; unknown scents, combinations of the sweet, the mordant and the subtle; a panorama, the sweep of the Bou Regreg River estuary and the ocean, the sultan's palace, palm trees and white European architecture, the coarser structures of the medina and, beyond that, African beehive huts.

Scattered figures below stilled, stood with bent heads and crossed arms, then prostrated themselves. The time was 'asr, halfway between noon and sundown, third of the five times daily for Muslim prayer, by command of God, even in war. We were struck by the serenity and calmness. The 800-year-old Hassan Tower, more correctly known as the Mosque of Hassan, rose nearly 100 feet above the seething nationalism and religiosity of 1955 Rabat, the capital city of French Morocco. This monument to victory over the Christians of Spain was unfinished to this day, mute testimony to eventual defeat. The Hassan Tower could be regarded as a monument to later defeat and internecine struggle; or, as a brick allegory for the determined climb by Moroccans towards freedom and independence from French control.

The horrifying 2001 attack on the Twin Towers in New York City by two packed airliners piloted by terrorists were examples of Muslim extremism. In the North Africa of the 1950s, Muslim frustration was still simple: they just wanted occupiers to get out and leave them alone. To make their point, they sometimes

Serenity of the Hassan Tower contrasted with the seething nationalism of French Morocco in the 1950s.

mobbed Europeans, and could be deadly cruel.

The Empire Builders were finally showing the world that there was nothing altruistic in colonialism, simply the economic impulse—some would say greed. Now that it cost too much to maintain empire, the colonialists were pulling out. But France was different. It learned lessons the hard way, with more turmoil, histrionics and grief than was prudent. France had exiled the tubby nationalist sultan, and replaced him with a bearded puppet. This divide-and-conquer manoeuvre had created factions, the deposed sultan now a political martyr with aroused and armed supporters involved in mounting waves of unrest—ambushes, riots, murders, massacres. Notices were posted about the "mood" in Rabat, and when cleared to visit, we had to be cautious about leaving the European section, or travelling in taxis with more than one Moroccan in the front seat.

In Algeria, the National Liberation Army, made up of roving bands of fellaghas, was so active in Mau Mau-type outrages that Vampire jets of the French air force frequently took off from Sale airfield, across the Bou Regreg River estuary from Rabat, to strafe and bomb the rebels in the adjacent country to the east. It was from this airfield that the Canadians now were flying their air-to-air gunnery exercises.

NEVER ABOVE THE SPEED OF SOUND, EH?

BEFORE LEAVING FOUR WING HOME base, we had got our shots by the station swimming pool. Two nurses on duty; two long lines of 410 officers and men; step forward, drop pants. It was like a furious dogfight—the nurses' hands, grasping needles, climbed to altitude, did a wingover and dived in a perfect quarter-attack, with deadly accuracy. "Ouch!"

"You may not know it," said Squadron Leader (major) Art Fisher in our Rabat-Sale flight room, "but the guns of the F-86 have never been fired above the speed of sound. So…"

He grinned, "…at the behest of Air Div, we're arming the kites—all six guns—and we'll do some supersonic air firing. One plane at a time to the range. Put her through Mach One, shoot, note what happens and give us a report. If there is anything to report."

The Raff's new Hawker Hunter had flamed out every time its guns were fired, and stiff upper lips had been hard to find in British air force circles until the problem was solved. No such problem with the Sabre, of course. Back in 1954, when sharp sonic booms rocked the urban peace in Leicester, or London, Ontario, the giant shock waves were easily blamed on one aircraft—the Sabre—the only operational aircraft at the time easily able to perform the feat. Sword drivers were cautioned that fighter pilots should be seen and not heard.

Never above the speed of sound, eh! Won't take long…if the kite doesn't flame out. Into the air and out above the ocean, heading east, 40,000 feet. Switches on, guns armed. Simply roll over and pull through. Straight down, machmeter needle approaching 1. The familiar hiccup—a drop of the left wing—as we go through the sound "barrier," leaving the wispy condensation ess, signature of the sound-breaker for all below to see (better not be anyone but the fishes below!). Trigger finger on the stick presses, the guns toc-a-toc-a-toc. At more than 700 knots an hour, black objects are still seen to whiz back from the nose and over the wing: the rubber gun port plugs pushed out by the first bullets. Th-th-that's all, folks!

Later that afternoon, we had peeled the tops of our G-suits back and rolled them down to the waist. Passing a football back and forth in the torrid sun. A sudden roar of engines, as an Air France DC-4, approaching to land, had been forced to overshoot when an RCAF Sabre was cleared onto the runway for take-off. The annoyed French airline pilot decided to emulate the Sabre pilots' tactics when they came in to land and had to overshoot: four Pratt and Whitney engines roaring at full power, the 40-passenger airplane flew just above the runway, slammed into a 45-degree bank to the left, climbed in a circle to circuit height, lowered its undercarriage and flaps again and came in to land, fairly bristling with indignation.

"What those passengers must have been thinking," someone muttered. "They probably were thinking guillotine," replied another. "That pilot better stay in the cockpit until the coast's clear."

You had to compensate for the heat when flying out of Rabat-Sale. Keep airspeed margins up ten knots or so. I forgot one afternoon and experienced some extra body heat problems. We usually had a Canadian in the control tower at the far end of the runway. Some pal of mine— Bud Venus maybe—was in the tower during my take-off run. I lifted off, cleaned up and, spur-of-the-moment, banked towards the tower. Level, nose pointed at the approaching tower. Flick! An incipient stall. The right wing drooped and righted itself. The Orenda was yowling its best. Another light flick. One could only hunch and hope. It wasn't my fault that the Sword behaved itself, built up speed and passed safely over the tower. Climbing away, a drinking song intruded: "…and they scraped him off the runway like a pound of strawberry jam, and he ain't gonna fly no more."

The rest of my time in Morocco my aircraft's nose was kept pointed down the runway bearing until well away from the field. Lift-offs were at about 160 knots, my landings hot.

MARVILLE, FRANCE

AIR DIVISION'S NO. 1 Fighter Wing at Marville, France, was a kind of plug in the middle of an invasion route that had been used countless times since before the Romans came to visit. The area between the Low Countries and Lorraine had been leased, pledged, exchanged, conquered and reconquered. Fire and sword became familiar through the centuries as it came under the rule of Germany, France, Burgundy, Austria, Spain and, finally, France again.

Marville base's ammunition dump was built atop a Roman road running from a Roman camp eight miles away, the original paving stones still visible. The famous battles of the Meuse, during the First World War were witnessed here. With the Belgian border only six miles north, the last bunkers of the Maginot Line were just outside RCAF Marville, and had engaged in battle as German panzers, with the aid of Stukas, did their famous end run and came down from the Ardennes into France, and on to Paris, in the early 1940s.

A sister Canadian airfield was located not too far to the southeast, No.2 Wing at Grostenquin. This was still in the storied old provincial region of Lorraine, a land of grain farms, cow pats, and coal mines that has been grabbed back and forth by France and Germany and nestles in old battle grounds, close to the Maginot Line and to the border of West Germany. France insisted that Canadian airfields would be built by Frenchmen. The result was disaster. When the first Canadians arrived at Grostenquin, in 1952, the site was a sea of rainwater and mud. Rubber boots were issued and were so necessary that they became allowable in the mess. The Canadian airmen nicknamed their station, Her Majesty's Canadian Ship Grostenquin. There were no buildings for accommodation, save for one all-ranks mess with sleeping quarters consisting of a bed, mattress, blanket, with showers down the hall. But the showers couldn't be used because of the water's high bacterial count. Personnel bathed from basins filled with chemically-treated water from a tank. Some enterprising airmen used wine for shaving.

No electricity yet was available. Propane was used for lighting and cooking for those few who obtained trailers to live in. Later, all plumbing and other services were supplied through above-ground plumbing. Grostenquin was unbelievable, when compared with the showcase base built later in Baden Soellingen by the Germans. Grostenquin was home base to 416 Black Lynx, 421 Red Indian, 430 Silver Falcon, all Sabre squadrons, and later, 423 Eagle (CF-100) Squadron.

Newly-arrived pilots at Marville, after driving and flying the area, commented on the scenery, the usual pleasant landscape augmented by huge piles of manure close to, and spilling onto, roads beside houses and villages. One Marville observer, an amateur historian, maintained it was a case of higher the pile, higher the esteem. "During the WWII occupation, one of the first of the German military commander's orders read to the effect that the years' accumulation of dung would be removed within 24 hours. It was removed. When the German occupation ceased, village air was again sweetened by growing piles of stable-cleanings, within 24 hours," said an article in One Wing's magazine, *Talepipe*.

Civilian flying and military flying in Europe were engaged in uneasy co-existence. Both used the same sky but seldom shared each other's flight plans. Usually, the military ignored commercial flying altogether; commercial pilots had no idea where the military pilots might be. Near-collisions were more common than generally realized. Many a military pilot averted disaster, through alertness or luck, and went home with no one else the wiser. One night flight out of Marville, in 1955, controllers assigned me the north-west quadrant. That meant flogging around north of Paris and over Belgium. Sifting along at 35 or 40 thousand feet, starlit sky, bright moon, sparkling Paris way to my left, then, an adrenalin alert: something ahead and to the right. A civilian airliner, long row of windows brightly lit, was flying across my path,

 CF-104s — this one in Marville — served in strike and recce roles with Air Division in Europe, under NATO, beginning in the early 1960s.

about 500 feet below. The pilot had no idea a plane was almost on top of him; I hadn't realized he was coming; if we were blips on anyone's screen, there was no reaction.

The next major European air war exercise was Carte Blanche, June 20 to 28, 1955, held in West Germany, north-east France and the Low Countries while 410 Squadron was stationed at Marville. Simulated atomic air raids were designed to "exercise all formations (air and ground) in the type of operations encountered in a major war." Civil and military aviation experts formed a committee to co-ordinate safety procedures and allow maximum freedom of training for the NATO forces. Cougar Squadron logged the most flying hours of the wing's three squadrons. The air attacks dropped 335 hypothetical nuclear bombs on military targets, killing 1.7 million people and wounding another 3.5 million. Though the war game's concept was unrealistic in area covered, and concentration of bombs, it prompted a wide public debate about NATO's nuclear strategy.

Warsaw Pact nations were busy with war games, too, during the half-century of the Cold War's threat of atomic annihilation. In November, 2005, Poland's government for the first time revealed top-secret documents including the Soviet bloc's take on a seven-day atomic holocaust between them and NATO. Maps showed red mushroom clouds blotting out Hamburg, Frankfurt, Stuttgart, Munich, Baden-Baden, Haarlem, Antwerp and Charleroi.

One Wing's—and 410 Squadron's—first Marville flying fatality occurred June 25, 1955, during Carte Blanche. Flying Officer Al McCallum was flying Combat Air Patrol in the early hours and, while chasing an RAF Meteor in an intense engagement, flew into trees. All of my air force training had been with McCallum, so it was an honour to lead his funeral procession to the RCAF section of the cemetery in Choloy, France. In 1956, our squadron lost another pilot in a flying accident. F/O Bob Roden was in a mid-air collision at 35,000 feet and ejected. Tragically, he was unable to free himself from his seat (there was no automatic release in those times) and fell to his death. Several of us were members of the funeral party that accompanied him to his homeland, England, for burial at Aldershot Military Cemetery.

We didn't make a show of death. There were no grief counsellors; we were stronger for that. If appropriate, we toasted our friend at the bar. More than 100 Sabre pilots are buried in cemeteries in England, France and Germany after guarding the freedoms and ideals for which Canada stands. Canadian military policy stated no bodies would be returned to Canada.

TO KILL WITH SKILL

IN ESSENCE, WE HAD been trained to kill with skill, and we couldn't help wondering what it would be like to actually exchange fire in anger, with all the risks to survival that would entail. Of course, we each would believe our chances were good. Through diligent, almost daily practice, we had developed some expertise. We never "wished" we could engage in actual combat. But it was like wondering about emergency ejection and parachuting to earth—we didn't want to but…what would it be like?

How would we have fared against MiGs? In the recently-ended Korean war, the MiG-15 had been a marginally better plane than the Sabre, although sustaining greater losses. The MiG-15 was faster than a Sabre above about 33,000 feet. It could climb at an attitude that would stall a Sabre. The MiGs could hold formation at 50,000 feet. Korean-war Sabres were sluggish 5,000 feet below that. The MiGs could turn a smaller circle.

But the Sabre was superior in a high-mach dive. The MiG started to snake badly above around mach .86—it wasn't a very good gun platform then. It was so unstable on a fore-and-aft plane that it went into a snap roll and spin when the pilot tried too tight a turn. The MiG had cannons. But not being a stable gun platform at high mach numbers, and with a relatively low rate of fire, their effectiveness was reduced. As well, it had a mechanical gunsight, vastly inferior to the Sabre's radar gunsight. All in all, while the MiG was a better flying machine, the Sabre was a better weapon. But the Canadian-built Sabres we were flying now, with Canadian-made Orenda engines, were a big improvement over the American Sabres used in the Korean war.

Some Russia-watchers in the West believed that Communist pilots, because of their political system, could not be "self-starters," show the initiative of NATO pilots. The political indoctrination, the time spent on discussions and lectures where officers tried to gain advantage over their peers, and the political reporting for dossiers all created an atmosphere of suspicion, of distrust. There was not time for much of a private life. And that is how the Soviet State wanted it. The fear that they could be accused of something, not only political deviation but even sabotage for an incident as minor as a blown nose wheel tire, blunted the pilots' initiative, some analysts wrote, and made them shy of responsibility.

The first air battles pitting jets against jets took place in Korea, in the early 1950s. At first, UN air superiority was threatened by Russian MiGs that were being supplied to the Chinese, who were backing the North Koreans. American F-80 Shooting Stars (forerunner of the T-33) and Panther jets, and Mustang propeller-driven fighters, were not up to the job. The first Sabres arrived in Korea in 1950, one of them flown by a Royal Canadian Air Force flight lieutenant, Omer Levesque (later one of my gunnery instructors), on exchange with the 4th Fighter Interceptor Wing of the USAF. He was the first Canadian to fly in all-jet air battles, and the first to make a kill.

Military historian Hugh A. Halliday wrote in an article in the Canadian Aviation Historical Society's journal, *CAHS*, that, "Then, the Sabres of the USAF won against MiG-15s with a kill ratio of 10 to 1, despite heavy odds. Among those Sabre pilots was a score of Canadians who contributed their share, shooting down at least nine MiGs and damaging many more." Twenty-two Canadians flew the Sabre in Korea, some of them flying Canadair Mark 2 Sabres, with American engines, more than 60 of which had been given by Canada to the Americans. One reason Canadian pilots flying in Europe were so well trained was the influence of Second World War pilots, particularly those who flew in Korea. I flew with, was trained or commanded by, Korean vets Bruce Fleming, Larry Spurr, Bob Carew, Duke Warren, Claude LaFrance, Bill Bliss, R.T.P. Davidson, Andy Lambrose, and Doug Lindsay.

The 10-to-1 kill ratio in Korea has often been challenged by cynics. But Halliday pointed out that, "Inexperienced Communist pilots were reluctant to take evasive action…

Their aircraft would be shot from under them, and they would then save their lives by ejecting. These low-calibre pilots were referred to as "jackpot flights." Bruce Fleming said in a post-flight report in Korea that poorly-trained North Korean pilots would sometimes eject before a shot was fired, or if their flight leader ejected.

Fleming told me of his weirdest Korean experience. He was flying near the Yalu River at high altitude. His eyes had been focused for long distance and so, as sometimes happened, missed something close by. He glanced to starboard and there, just a few wingspans away, was a MiG-15, headed in the same direction. The MiG pilot appeared to be just as startled as Fleming. Fleming popped his speed brakes and chopped his throttle; so did the MiG pilot. Still, neither had achieved an advantage. Fleming dumped flap; so did the MiG pilot. They were eye-balling each other, sensitive to the danger of making a wrong move. "I could see he wore an orange helmet under his hard hat. I'm sure I would have recognized him if he showed up at the mess that night," chuckled Fleming.

Finally, Fleming raised his right arm and waved, tentatively. The other pilot returned the gesture. Then carefully, very carefully, each eased away from the other and parted company.

ZULU SCRAMBLES

A JANUARY, 1955, Zulu Alert at Zweibrucken illustrated how innovative our ground crews could be in keeping us in the air and safe. The first morning—4 a.m., below freezing, miserable wind—we piled into the Alert tent, then went out to check the aircraft. A thin layer of snow and ice coated the wings—a condition that could adversely affect flying stability of the aircraft. The flight commander consulted with the ground crew and then told us—already flaked out on folding cots, one ear listening for the field telephone's jangle—to go out and follow the makeshift de-icing procedure developed during earlier winter alerts. Four aircraft at a time were started and taxied to an open area. Two Sabres were parked tailpipe-to-tailpipe, engines idling, their exhausts a little farther apart than the 37-foot wingspan of the aircraft. The other two planes taxied slowly between them, the exhaust heat melting the snow and ice, the buffeting shaking the pilots. Then the procedure was reversed. Finished, the aircraft moved back to the flight line, were shut down and their fuel tanks topped up. Then four more went out. When all were de-iced, four of the aircraft were left standing with energizers plugged in and humming, parachutes and helmets in the cockpits, ready for hasty start and take-off. The other four were on immediate standby.

Possibly my most questionable take-off on a Zulu scramble occurred at Zweibrucken. This particular morning, we lounged inside the Alert tent knowing that the energizers (to assist engine starting) were plugged in and purring, crash helmets were perched on top of the windscreens, and parachutes inside the cockpit on top of the seat, straps laid carefully to each side. Before the scramble order rudely brought me back to earth, I had been musing on the influence of machines—such as the ones waiting outside—and why it was so many advances in technology came during wars between those the mechanical wonders were designed to benefit?

At the whir-r-r of the field telephone we were standing, poised, and after our leader picked up the receiver and repeated the word "scramble," the four of us were running the hundred feet to the kites, donning parachutes and helmets, strapping in, doing up, turning on. Flip the engine master switch to ON, hold the battery-starter switch momentarily at the STARTER position, then move it to BATTERY. At three percent rpm, throttle outboard to energize booster pumps and start ignition. At six percent, throttle advanced to obtain fuel flow of from 500 to 700 pounds per hour....Damn! False start. Engine speed had reached nine percent but no ignition. Hit the PUSH TO STOP STARTER button. Bursts of power beside me as the other Sabres moved out and careered towards the end of the runway, no more than 50 seconds from the order to scramble.

Throttle off. Supposed to wait three minutes before trying another start Drain fuel, cool starter...too long. I began the start procedure again as the other three aircraft reached the runway's end.

"Coming Four?" Leader's voice. The other three began their take-off roll. "Roger." Aside from a slightly

The author's Sabre above the Marville runways blends with farm textures. The base's ammunition dump was built atop a Roman road running from a Roman camp eight miles away. *Photo: Wilf Thorne*

elevated temperature, the start was successful this time. I goosed the engine and roared along the curving taxi strip towards the runway, skipping a little while negotiating the turn. They were lifting off as my Sabre skidded onto the strip and the throttle was slammed full forward. My speed built rapidly. Needing to catch up, I left the aircraft on the concrete until it was doing 180 knots, lifted the nose wheel, then reached forward to the left and raised the undercarriage control lever. While still in contact with the runway, the main wheels entered their retraction cycle. The D-doors dropped open, the wheels folded up. The aircraft settled slightly. Oh, oh! A momentary rasping sound as the D-doors scraped along the runway. That meant about one thirty-second of an inch of aluminum was now missing from the D-doors. Didn't intend that, but no problem.

Had the controller in the tower poised a finger over the crash bell button as he saw, in the dingy winter light, sparks beneath the roaring Sabre before it was airborne and everything seemed okay? Actually, I enjoyed the sensation and repeated that take-off procedure several times in future, taking care not to drag my D-doors.

A check forward on the stick, holding the straining Sabre a couple of feet above the runway to gain speed more quickly. Terrain at the runway's end sloped down into a shallow valley that stretched to the right. The others were climbing slowly to starboard. I banked that way but held the aircraft down, settling into the valley, just clearing trees. My speed built up so rapidly that soon my aircraft was beneath, but far below, my climbing buddies. I eased back on the stick, began a slow climb.

You never know when Life will unwrap a revelation for you. Several hundred feet of altitude were indicated when a red light on the upper right of my instrument panel demanded attention—the aft warning light. No sweat! The one above it, the forward warning light, was the one we didn't like: it meant you had to eject immediately; this one required the pilot to retard the throttle and see if the light would go out. If it did, he was to ease the throttle ahead again until all was functioning smoothly. I did this. Everything was fine. The engine pulsed strongly, the light was out.

Pulling up to the rest of the formation now, feeling satisfied with my calm reaction to the potential problem, and how easily it had been fixed, I pressed the transmitter button to inform the leader that Number Four was "coming on board." "R-R-Red L-Lead-Leader…" I was mortified.

Despite my feeling of easy control throughout the emergency, stress had obviously gotten to me.

Most months, the total number of flying hours for 410 Squadron ranged from 450 to 550 hours. Each pilot's missions, and times, were listed on a flight room board. One of our most productive months at Marville was September, 1955, when, with 24 aircraft to strength, we totalled 705 hours and 50 minutes. Many of our flights were without auxiliary tanks, and those usually lasted no longer than 40 to 55 minutes.

For that September, my logbook shows 50 flights for a total of 43 hours and 10 minutes. Towards month's end, we obviously had flown with auxiliary fuel tanks, as times were as much as an hour and 55 minutes. Several days we were in the air three and four times. There was a great variety in these missions. One was to fly as a "bouncer" and attack a formation of aircraft; on another, I recorded a normal hydraulic failure due to a wing-line break, getting back to base on the aircraft's auxiliary hydraulic system. Other flights included RAF exercises; high battle formation; test flights; aerobatics; navigation trips to Luxembourg, Cherbourg and Amsterdam; tailchases; GCAs; two-versus-twos; dropping leaflets on an American base inviting their pilots to a ball at our mess.

Starting in June, 1956, our Mark 5 Sabres gradually were ferried to Langar, England, for eventual shipment back to Canada, as 410 Squadron re-equipped with Mark 6s. Then, in October, we found that our squadron would be disbanded to make room for 445 (Wolverine) Squadron, with its CF-100 twin-jet Canuck all-weather aircraft. The mad scramble began to get as many flights in our beloved Sabres before The End. That September, 410 pilots flew a record 756:10 hours.

One late summer afternoon, at about 45,000 feet, we noticed what appeared to be a phantom aircraft flying in the same direction, due east, but at an altitude we couldn't believe—at least 65,000 or 70,000 feet, we estimated. It was a hazy image we stared at, lightish grey with no apparent markings. We pushed our Mark 6s into a climb, passed through 51,000 feet, but gave up any further attempt to reach it. The mystery plane had a high aspect ratio, very long, narrow wings, glider-like—103 feet wingspan we learned much later. Over the next while, we heard reports of other pilots from other wings seeing the aircraft.

We had spotted a U-2 spy plane, first flown a year earlier, a single-seat, single–engine reconnaissance aircraft, carrying cameras and sensors. It was flying regular sorties in the stratosphere over the Soviet Union. E.J. Chenier, a former Yellowjack controller, wrote in a 2004 letter to Airforce magazine: "I recall on many occasions, we would pick up an unknown and would scramble our "Zulu" fighters to intercept. Our F-86 flyboys would get a tally-ho on the unknown which was a U-2 up in the 70,000 feet-plus range and by this time, the Americans would call us off and say it was a "classified mission", and we were told to break off."

Russian MiGs and missiles couldn't reach the U-2 until 1960, when a U.S. pilot, Gary Powers, was shot down, baled out over Russia, and was taken prisoner, creating high international tensions. It was a U-2 that photographed Russian missiles in Cuba in October, 1962. Production ended in 1968, but they still were flying in the New Millenium, and were reportedly used operationally over Iraq in 2002.

Many Sabre pilots, in 1956, were anticipating the day they could climb into the mighty Avro Arrow and show the world Canadian inventive genius. Tales came back from the Arrow plant about these sleek, mach-busting aircraft, that the leading edges of their wings were so sharp they were covered with felt pads while in the hangar so maintenance people would not sustain cuts.

In October, 1956, many 410 pilots were absorbed by 439 and 441 squadrons. Some of us packed and left Marville, boarded the Homeric (my wife and I, and now two children, as well) and sailed across the Atlantic back to dear old Canada where six months as a flying instructor in T-33s awaited me, before I re-donned mufti and returned to my first call, the never-predictable, cusp-of-events life of newspaper work.

The magnificent Sabres managed to continue without me, but the last RCAF Sabre flight took place in December, 1968, at CFB Chatham, New Brunswick, the spawning ground of so many Sabre pilots. •

Photographer Vic Johnson twists around for a dramatic shot of his pilot and three other CF-18 Hornets in then West Germany in the 1980s, when the aircraft began their NATO and NORAD roles. The first CF-18s flew with 410 (Operational Training Unit) Squadron at CFB Cold Lake, Alberta, in 1982.

CHAPTER SEVEN

A TIME OF TRANSITION

WE HAD BOUNCED IN Canadian UNFICYP (UN Forces in Cyprus) vehicles north from Nicosia, penetrating the Turk-Cypriot enclave along with Princess Patricia Canadian Light Infantry personnel to St. Hilarion Castle. There, the star and crescent of Turkey snapped determinedly above the gatehouse just beneath Prince John's Tower, and above the soldiers' firing range which so long ago was a jousting ground for King Richard the Lion-Heart's crusading knights.

This was 1972, and the Turk-Cypriots were at the mercy of their former Greek-Cypriot countrymen. Electricity and other necessities could be cut off at the whim of Greeks. Left alone, the Greek and Turk Cypriots had lived together in friendly co-operation. But the politicians, those people so often of little sense and large egos, fouled things up. The current misery began in the early Fifties when Greek-Cypriot politicians decided they wanted union with Greece. When a Greek military junta moved to take over the island in 1974, Turkey understandably invaded from the north. That split the island.

Before leaving Canada, I had read Lawrence Durrell's impressionistic study *Bitter Lemons*. He had renovated an old house in Bellapaix, near the Tree of Idleness by a cafe on the north slopes above Kyrenia, and lived there while the troubles were brewing in the early Fifties. An endearing local character was a wine seller, Clito.

I wanted to find Durrell's house. I had thought about it several times during the RCAF Boeing 707 flight to the eastern Mediterranean, a scheduled service for its NATO units. Trudging up the slopes into a sorcerer's world, my feet crunched pomegranates, lemons and oranges. Falling time, November. With a guide I climbed ever higher up the foothills in the sun-split shade of orchards, mulberry, carob and cypress trees. There! My young point man

VIA AIR MAIL
CORREO AEREO
PAR AVION

View from Prince John's tower at St. Hilarion Castle, in the Turk-Cypriot enclave of northern Cyprus, is of a soldiers' firing range, once a jousting ground for King Richard the Lion-Heart's crusading knights.

Gleanthis Thalassinos indicated at last. From the street, Durrell's blue-and-white house could have been a small closed-up factory. But front windows looked down on natural splendour—olive and orange trees, Kyrenia's harbour, and out onto the glazed Mediterranean and across to Turkey. We stood on the far side of the tilted street, lolling kind of stunned against a concrete aqueduct, its roaring white flow tumbling from the higher mountains.

Later, our military vehicle rolled into the streets of Kyrenia below, and a sign on the corner of a neglected building caught my eye. Enamelled metal, white on blue—Clito's Bar. Above and hard against it, a newer sign added after fame was bestowed: "Bitter Lemons". I was dazzled. We all tumbled out, climbed worn wooden steps and in to dark and sombre cool, a cavern. Glasses filled with wine and ouzo, the group of writers and TV cameramen from Winnipeg and Northwest Ontario clumped downstairs to a basement of small round tables, chairs and an ancient Wurlitzer with spritely Greek music which soon had Major Dent, our air force public relations military guide, gyrating.

Upstairs, behind the scarred wooden bar backed by a double-tiered row of signed wine barrels, dripping spigots, the elderly barkeep sat. Jacket and pants of black fustian, sweat-stained open-necked shirt and a pleasant if vacant expression. "More Coca-Cola?" he joked.

I mounted a stool: "Is Clito still here?"

"I am Clito," tone implying pride, pleasure that he could impress me so. I had been greatly taken with Durrell's impressions of Clito and was momentarily speechless. *He stood behind his own bar with a faint and preoccupied kindness graven on his thin face...* But now, his eyes were focused straight ahead, unwavering, examining images only his mind could see. Even while drawing wine and placing it on the counter.

Casual talk. Then: "In Bitter Lemons, you weren't...blind."

"No." He still smiled, stared.

"How long?"

His hand floated forward, found the cigarette smouldering in an ashtray, put it to his lips and he dragged. "Two years, just before Christmas."

"What happened?"

He shook his head. "It was a mistake. Two Canadian sergeants with the Black Watch. They were having a good time. They beat me up." He paused. "But they were nice boys." His head shook slightly, still hard to believe. I was stunned. This Clito who had figured so large in my reader's imagination subjected to such gross treatment. My own countrymen.

"Is that all?"

"The Canadian government awarded me five thousand Cypriot pounds." His tone suggested fair compensation. I couldn't imagine that it was. Loss of sight. Out came my battered copy of Bitter Lemons.

 Four CF-18 Hornets from Canadian air force base Baden-Soellingen, in then West Germany, soar above Bavarian foothills. Three squadrons—409, 439 and 421—flew Hornets starting in the 1980s. In 1991, 26 German-based Canadian Hornets went to Dohar, Qatar, to participate in the Gulf War. *Photo: Vic Johnson*

Like a literary groupie, I asked Clito to sign it. Slowly, carefully, he did, twice, once with the Greek spelling, a K. Done by memory and feel, surprisingly legible. I thanked Clito and rejoined the bunch downstairs. It wasn't the same. Later, back in Canada, the judge advocate general's office in Ottawa confirmed Clito's story.

SEVERE TURBULENCE FOR THE ARMED FORCES

MORE THAN FIVE DECADES AGO, the blue NATO Star flag was paraded for the first time outside of its headquarters at Brussels—on the sunlit tarmac of the Royal Canadian Air Force station at Winnipeg's airport, in symbolic testament to Canada's heavyweight support of the international defence group. Those complaining that Canada is asked to take on too heavy a NATO duty in Afghanistan shouldn't be surprised that the country has several times punched heavier than its weight. But they may be surprised to find that PM Stephen Harper isn't punching as hard as earlier prime ministers.

Chinese Communist aggression in Korea, in 1950, engaged the UN and spurred the one-year-old NATO into action, and soon Canada was heavily involved with both organizations. While sending soldiers and jet pilots into battle in Korea, this country once again started training aircrew for the free world and, as well, set up a huge defence establishment in Europe. Canada was eager to work with NATO. Several years as a defence partner with the United States had made Canadian politicians welcome more companions-in-arms. The saying going the rounds in Ottawa, was, "Twelve in the bed means no rape."

From 1949 (when NATO was founded) to 1956, PM Louis St. Laurent spent 6.5 percent of GDP on defence. Under John Diefenbaker, this fell to 5.4 percent, then down to 3.8 under Lester Pearson. Under Pierre Trudeau, spending fell to 2.1 percent, then to 2 percent with Brian Mulroney, 1.3 percent from Jean Chretien, 1.2 percent under Paul Martin. Despite billions of dollars earmarked for defence by Stephen Harper, the percentage of GDP is still at 1.2 percent. Actually, Harper fits in neatly with the four previous under-achievers, pushing all the same vote-getting buttons.

AIR FORCE TRADITION

IN 1968, 47 YEARS AFTER separation of the air force from the army, Prime Minister Lester Pearson's Liberal defence minister, Paul Hellyer, re-merged the two military branches (and the navy) into the Canadian Forces, and the RCAF ceased to exist. The Pearson-Hellyer political tactic removed the "Royal" from the air force title. It discarded the traditional Royal Air Force blue-grey uniforms, and became more American, even switching to the America salute. The RCAF/RAF motto, *per ardua ad astra*—"through adversity to the stars"—was dropped. All this and the rank structure were dumped by politicians to satisfy those who rejected early British connections. But when Pearson's successor, Pierre Trudeau, tried to remove "Royal" from the RCMP, Canadians erupted in fury, and the prime minister bowed to the wishes of the people.

Both the RCAF and the RCN opposed the politically-inspired move, but were over-ruled. The merger was announced as a brilliant political concept that other nations would rush to follow—but none has. After a period of dissatisfied existence with no distinction between previously-proud military structures, the homogenized dark-green uniforms for all service branches was dropped, in 1988, for a return to more traditional hues. This seemed to affirm Field Marshal Lord Wavell's advice

The blue NATO Star flag (left of centre) was paraded in the 1950s for the first time outside of Brussels—on the sunlit tarmac of RCAF Station Winnipeg—in symbolic testament to Canada's heavyweight support of the international defence group.

in a 1949 talk in Montreal: "This modern tendency to scorn and ignore tradition and to sacrifice it to administrative convenience is one that wise men will resist in all branches of life, but more especially in our military life."

And, in 2006, retired brigadier Claude Thibault wrote: "The CF-5 era began in the very late '60s and ended in the mid '90s. This will be remembered by some as a period of great uncertainty in command, control and direction. Our focus was certainly unclear but not our enthusiasm." Although the title "RCAF" had been eliminated, its ghost still flies: April 1 still is observed as the anniversary of the birth of Canada's air force. In 1975, Canadian Forces Air Command (AIRCOM) was created, and became responsible for most aviation units. AIRCOM preserves many traditions of the RCAF, such as the RCAF tartan and the command march. In 1993, air force formations called wings were reintroduced within AIRCOM, echoing the structure of the RCAF thirty years previously.

Defence-spending cut-backs took on a life of their own a year after unification of the armed forces. Prime Minister Pierre Trudeau, shrugging at consultation with NATO, cut Canada's contribution in half, forgetting about the link between defence spending and trade between nations. Wrote historian Desmond Morton, in *A Military History of Canada*: "In 1976, after years of embarrassing indecision, Ottawa finally purchased 128 German-made Leopard tanks for the NATO combat group. The order sweetened European trade negotiations and helped secure Canada a seat at the European Security Conference at Helsinki. It hardly seemed to matter that the tanks themselves were already outdated by German, British and Soviet standards."

With the forces embarrassingly weak, Trudeau's response was woeful. After years of political infighting and commercial complaints, he spent $4 billion for 137 fighters to replace the worn-out Voodoos and Starfighters. Six contractors had bid for the fighter contract, but four of these had been eliminated because their aircraft cost too much. The winner was the best of two somewhat dubious choices, the McDonnell-Douglas F-18D Hornet. The loser, General Dynamics, put some spark into the 1980 Quebec referendum campaign by claiming Ottawa had favoured an Ontario branch plant. Lost in the murk of all this was the fact that an early loser, the Panavia Tornado, was the closest to what Canada needed, and would have involved Canada with Britain, Germany and Italy. So Trudeau had ignored NATO and limited the choice to American aircraft.

Paul Hellyer had promised that integration of National Defence headquarters would introduce an age of efficiency, co-operation and economy. Events have done little to prove him right. In 1972, Lockheed offered twenty-three long-range patrol aircraft for $300 million; by 1978, the price had tripled for five fewer planes, Morton wrote. "Other equipment purchases were as contorted by political and business pressures. Civilianization, two conflicting White Papers, and a grandly titled, 'Defence Structure Review' in 1974 had not helped. Perhaps the real problem lay less with the Department of National Defence and its constant procession of ministers than with the concentration of power under a man more brilliant in debate than in decision-making. Ministers could extract promises of seven and even twelve per cent additions to defence budgets but 'freezes,' 'squeezes,' and 'cutbacks' regularly delayed purchasing decisions and added to their costs."

Morton explained "civilianization" this way: Trudeau and his Privy Council Office decided that generals and admirals should not make important decisions. Instead, changes affecting the nation would be made by civilians. "Civilianization" meant that in 1972, service heads of branches at National Defence Headquarters were re-titled assistant deputy ministers and, in some cases, replaced by civil servants. "In the prevailing mood, military leadership and perhaps even war itself, were obsolete."

GOING HOLLYWOOD

THE CANADIAN AIR FORCE linked arms with the entertainment business early in 2008 when CTV's Discovery Channel aired an eight-part series on the training of fighter pilots. The slick program had the poignancy of the attention-deprived crying, "Look at me, look at me!" After years of being deprived, the air force wanted attention.

 Canada has its foot in the door for a slick NATO jet, the Martin F-35 Joint Strike Fighter, the Lightning, contributing money for design, building and testing.

Hollywood's myth-making has made use of the U.S. military's expensive hardware and technology practically since the early talkies, but Jetstream was the first such venture for the air force in this country, a $35-million slick production that had front-line CF-18 jet fighters and pilots as its stars. Filming mostly was done at 4 Wing, the internationally-envied vast air weapons range at CFB Cold Lake, straddling the Alberta/Saskatchewan boundary.

But, well done and interesting as it was, it came off as yet another version of the ubiquitous survival-program format: which of the pilots would be the first to be kicked off the training program; how many would make it to the end? While the series made it appear that our fighter pilots have a great future, in reality they are closer to being cast-offs, the only big money for the air force being spent so far on heavy-lifting aircraft to support the army—commendable, but simply supporting the army was the role of air forces in 1914. The old but upgraded CF-18 operational force has been greatly reduced (less than half the original 125 aircraft still are in squadron service) and there is less for new fighter pilots to do, and not enough money to pay for very many of them.

The students in the production all had to go to Toronto and undergo the whirligig delights of a centrifuge, to see how much "G" force they could withstand, before losing consciousness, when blood is squeezed from their brains and hearts into lower extremities. One pilot was "CTd", ceased training, because, physically, he didn't measure up to the requirements.

A few years ago, a high-ranking Canadian air force officer told a group in Winnipeg that there were not enough qualified pilots to man available fighter planes. To save money, the air force had pared down its CF-18 fleet to 80 aircraft; there were supposed to be 68 pilots available to fly what was left, but only 63 pilots at that time were on strength. That was the pilot strength of just one of our overseas wings at the height of the Cold War.

The refrain most commonly heard in recent years from disgruntled air force officers is that their once noble organization is in danger of extinction in a period of army-centric defence thinking. They decry reduction of maritime air patrol capability to a fraction of what it was (Newfoundland's Provincial Airlines has contracted since 1989 to do maritime patrol flights. It replaced Canadian Forces aircraft on fisheries patrols, as well as providing gathered surveillance data to the department of national defence). They complain that long-overdue updates to the CF-18 fleet are followed by reduced flying hours, so that core capabilities such as low-level operations are abandoned. Supply capability is

G-SUITS

Pilots of all world air forces have been wearing G-suits of some kind or other since the Second World War. I wonder if any of the participants in the Jetstream program, or the viewers, were aware that both the basic G-suit and centrifuge were invented in the 1940s by a Canadian aviation medicine specialist, Dr. Wilbur Franks, a University of Toronto graduate? Because of the war, the G-suit information was freely distributed to Canada's allies. Eventually, Franks was named to the Canadian aviation hall-of-fame, but at the time, official Ottawa had ignored him. Aviation historian Peter Allen lamented, "I think that's part of the Canadian apathy. We're great at hiding our candles." However, Americans heaped honours on our ignored Canadian for his work, because they recognized and respected achievement.

contracted to civilians. T-33 and fleets were retired to pay civilians the same money for a fraction of the capability and deprived the army, navy and air force of realistic combat service support.

"We reduce personnel training standards in many important areas and impose other standards that have nothing to do with combat capability...We do all this and call ourselves more capable. We're fooling nobody and our service personnel, country and allies deserve better," complained retired air force lieutenant-colonel, Laurie Hawn, later Conservative MP and parliamentary secretary to the defence minister.

Canadian forces in the field need support from a revitalized air force with its own up-to-date helicopter transports and gunships, backed up by heavy-fire jet fighter-bombers, a capacity in which the air force is woefully unequipped today. Although it's not locked in, Canada has its foot in the door for the needed jet fighter-bomber—NATO allies are assisting in development of a joint American-British $300 billion project, the Lockheed-Martin F-35 Joint Strike Fighter. Italy, The Netherlands, Canada, Norway, Denmark, Australia and Turkey have formally joined and contributed money toward the program. They are not just lined up to buy the airplanes, but are participating in the design, building, and testing of the F-35. Four proto-types have been built and flown.

These partners are either NATO countries or close U.S. allies, and peacekeeping and war fighting more recently have been done by coalitions. With the F-35, allies can all fly the same airplane. The aero industry in each of these countries expects economic benefits (by 2006, 54 Canadian companies, universities, and research institutions had won 154 JSF-related contracts, worth $157 million. In September, 2008, Winnipeg's Bristol Aerospace, already making structural components, announced $120 million in re-tooling for further expected F-35 contracts. The federal government's Strategic Aerospace and Defence Initiative made $44.3 million available to Bristol). The F-35 can almost be regarded as an industrial policy, said *Aviation Week*. It was the ministry of trade — not defence — that cast the deciding vote to join the Joint Strike Fighter team. The aircraft was named Lightning after the wartime Lockheed P-38 Lightning, and the English Electric Lightning. Current plans call for the U.S. and U.K. to purchase approximately 2,600 aircraft, with potential sales of more than 2,000, starting around 2013, to the other NATO participants, including 60 for Canada.

Sensors provide the pilot with far more precise search and targeting capabilities than exist in today's attack fighters. The F-35 is also equipped with an infra-red search and track (IRST) system for air-to-air combat, and advanced air-to-ground combat features. There is a speech recognition system that detects a pilot's spoken commands and operates various systems without the need of pressing buttons or flipping switches. The design emphasizes stealth.

The air force will buy its own variants of the F-35, after spending, up to now, $2 billion on the development program. The plan is to form two attack-helicopter squadrons to escort Chinook transports and to back up the army. The F-35s will be supported by two aerial tankers. Four heavy lift C-17s (done) are for the army. A Canadian battalion could be flown just about anywhere, at short notice, by medium lift transports, probably Hercules C-130Js. There will also be new fleets of maritime helicopters (which are about to arrive) and fixed-wing search and rescue aircraft to go with the Cormorants already flying. Strategic unmanned aerial vehicles (UAVs) will be used for long-range surveillance of Canada's coasts and the Arctic, and tactical UAVs will be used by army formations.

What about the top of the world, the Arctic region that is becoming more and more like the old Wild West, with blurred sovereignty lines and lawless venturing? A race already is on to establish claiming rights for minerals, oil and even fresh water. To make other countries respect its presence in regions that Canada claims, it must have an increased military presence, with muscle. The Canadian military must become much like the North-West Mounted Police presence in the early West. Canada already has announced a significant increase in its military presence, including up to eight more navy ships and a deep-water refuelling facility for naval vessels. Also needed are at least a couple of major airfields for CF-35 squadrons (or whatever is purchased) and for top-line helicopters. Although armed conflict is not expected, Canada needs at least to be able to counter any vessels engaged in illegal fishing or unsafe shipping.

Denying adequate funding to the defence department and asking it to muddle through somehow, cannot be intelligent policy. Instead of being a "bothersome afterthought," shouldn't national defence become a priority concern of our political leaders? The basis of Canada's almost constant refusal to put dollars into defence, preferring to spend them on social comfort and votes, is the fact that, because of our geographical position—nestled against the strongest nation in the world—there has not been a pressing need to defend ourselves and, in that respect, we have become spoiled silly. Even our contract with the North American Aerospace Defence Command (NORAD) is limply supported by our political rulers, perhaps because they reason that the U.S. needs us (a place to park their defences) more than we imagine we need them.

And it is a conundrum. We shouldn't refuse, as the Paul Martin Liberal government did in 2005, to participate in ballistic missile defence, which was a serious blow to the concept of mutual protection against a common threat, the basis of NORAD for the last 50 years. But, it is also difficult for some Canadians to differentiate between the close collaboration of Canada and the U.S. that is required, and the danger of excessive integration and loss of sovereignty. (But, if we wanted to protect our sovereignty from the U.S., and in the Arctic as well, then why didn't we build the warning systems in the north, instead of just providing the sites? We've had rocket technology for almost as long as there's been a NATO, so why didn't we develop our own missile defence?) And that seems to be the root of our selfish approach that refuses to spend enough on our own defence mechanisms, let alone mutual protection. The facts suggest that, even if we did agree to support NORAD to the maximum, we aren't providing our own pilot and fighter plane muscle to back it up.

One senior defence analyst, Dr. Julian Lindley-French, has written in *Reconnecting Canada to the World—Via Europe*: "The armed forces are one of the few national institutions that represent the interests of all citizens, and stand ready to apply a sense of unity across a far-flung dominion…Defence recovery will only occur when Canadians recognize the vital role armed forces play in support of their well-being."

Skeptical critics have claimed that the Conservatives have been scaling back commitments they made during the last election campaign. Prime

⤞ Squadrons of Canadian CF-100 Canuck all-weather fighters were long the country's main air defence against long-range attackers, also serving with NATO in Europe.

Minister Stephen Harper has repeated the pledges, without mentioning deadlines for action. A defence department performance report, released in November 2007, concluded the government had not allotted enough money to meet the target set in 2006. Harper has promised annual defence spending of $30 billion, by 2028. The Opposition Liberals argued that the Conservatives claimed they had rebuilt Canada's defences, but they would have to spend much more than $30 billion over 20 years to get the military back in shape. Canada's annual defence budget should be upped to $35 billion, from $18 billion, by 2011-12, the Senate defence committee insisted.

Regarding Harper's position, one columnist tsk-tsked: "Yet this let's-pretend document salivates for tanks and barely mentions fixed-wing aircraft, surrendering the job of patrolling 71,261 kilometres of coastline, and our vast interior, to the duct-taped-together Buffalo and Aurora fleet for another decade."

There is a parallel between under-funding the military and under-funding Canadian Olympic sports programs. Like the air force, sports suffered funding cuts in the 1980s. "…winning became a dirty word, There was mediocrity. It was okay just to participate," said Alex Baumann, Olympian and head of the Canadian Olympic Committee's Road to Excellence program, in a *Maclean's* interview. "I think we should strive to be the best in the world. It doesn't mean we win at all costs but I don't think sport is any different than academics, or art or business—we should strive to be the best in the world."

If we wish to be a strong and proud nation, shouldn't we be less reluctant to foot the bill—financially and intellectually? •

 The real tragedy of the Arrow cancellation was the Iroquois Engine, most powerful in the world, which was designed, tested and manufactured by Orenda, and was at least 10 years ahead of its time.
Photo: Courtesy Airforce Magazine

CHAPTER EIGHT

AEROSPACE: A POWERFUL CANADIAN SYMBOL

THEY CALLED IT 'FIFTY YEARS FLY-PAST' and on a sunny June day in 1954, air enthusiasts flocked to the single, mile-long grass airstrip known as Coventry Civic Aerodrome (RAF Baginton) to watch the past mingle with the future. Reciprocating engines mounted on canvas-covered stick-planes hiccuped in competition with the throaty howl of jet-propelled, stressed-cockpit jet aircraft in one of the most unusual air displays to that time.

Flying Officer Bill Johnson and I joined eager English civilian watchers downing chips and squash, who were wowed by some of the latest Royal Navy and our Canadian jet aircraft from No. 1 (Fighter) Wing, North Luffenham, Rutland, performing at the Warwickshire airfield, built in 1935 three miles southeast of Coventry. On the other hand, the military jet pilots at the show were wide-eyed at the old-time airplanes. Examining them on the ground, they doubted that they still were capable of flight.

But due to care and excellent maintenance—and some might have argued perverseness on the part of the airplanes—many responded to their pilots and lifted off the grass. The Humber Bleriot XI, a reproduction of the famous type which Louis Bleriot flew to achieve the first cross-channel flight on July 25, 1909, defied crowd expectations by gaining altitude until it was at least 12 feet above ground after a burst from its 22 h.p. Anzani engine. Not doing much better, but still a thrilling sight, was the puttering of the 50 h.p. Gnome powering one of the first practicable British airplanes, the Blackburn Monoplane, designed in 1912. Other aircraft taking part in the 1954 Coventry flying and static air show included: RAF English Electric Canberras; an Avro Lancaster; a Hawker Sea Hawk; biplanes such as the Hawker Tom Tit, Fairey Swordfish, Hawker Hart, Sopwith Pup; a de Havilland Dragon Rapide; a Hawker Cygnet; a Gloster Meteor; a Canadair Sabre piloted by 441 Squadron's F/L Dean Kelly; and the RCAF aerobatic team of four Sabre jets from North Luffenham.

Most of the gawking fans on the ground were thrilled to see the twisting tell-tale S of white condensation vapour writ on the high blue when Kelly's Sabre went through the sound barrier, and by the reverberating triple boom when it arrived. What they didn't know was that the Canadian pilot's headrest had come free when he rolled the plane onto its back for the dive towards the airdrome. Always relaxed, Kelly finished the dramatic manoeuvre, and a tricky low-level display, with the piece of aircraft wedged under his arm.

A Royal Navy Sikorsky helicopter rested on the grass behind a rope fence and we asked the pilot if he would take two Canadian fighter pilots for a flip (my secondary duty at 1 Wing was editor of the wing magazine, *Talepipe*, and I was taking pictures). Soon we were strapped in beneath the whirling rotors, ready for our first helicopter flight. The take-off was so smooth and fast that we thought the ground had simply dropped away. We spent most of the flight hovering along the ruins of Coventry Cathedral, burned to the ground the night of October 14, 1940, following a devastating Luftwaffe air raid.

The Armstrong Whitworth Aircraft factory, on the airdrome's doorstep, turned out Whitley, Lancaster and Lincoln bombers, and later, for Gloster and the Hawker Siddeley Group, Sea Hawks, Hunters and Meteors. Today, Coventry is known for its Midland Air Museum's Sir Frank Whittle Jet Heritage Centre,

⇒

These two British relics of bygone flights, a Sopwith Pup and a Fairey Swordfish, were showing their stuff at an air show near Coventry, very close to the source of Canada's jet engine expertise that powered the great CF-105 Arrow (Opposite).

which bases two main themes: The Story of the Jet, and Wings Over Coventry, showcasing Armstrong Whitworth and Gloster aircraft. The Whittle Centre has close links to Winnipeg and Canada's great Orenda jet engine story.

GIANT LEAP BECAME A FUMBLE

IN 2009, BY THE END of the first century of flight, Canada had left no doubt that it was, and is, a heavy-hitter in the aviation world—it was a huge contributor to keeping the world free during the Cold War, an effort surpassed only by a magnificent effort in the Second World War; and in civil aviation, Canada developed one of the most successful aircraft manufacturing companies in the world. The nation can be expected to hit even harder in the next 100 years of aviation.

Canada was able to support NATO during the tense 1950s with planes and pilots that ranked best in the world. But then, when it had been ambushed and stripped of the ability to launch the futuristic Arrow fighter plane powered by the world's best engine (or even to use the expertise acquired for future development), government authorities turned their backs and forgot all about those things.

Wartime Winnipeg provides a fuller picture of the development of the world's best jet aircraft engine of that time:

The unfathomable shriek of a jet engine from behind a Stevenson Field hangar in 1944 Winnipeg should have been the foreshadowing of a giant leap forward for Canadian greatness in aviation. But the leap was fouled by a trip-up, and became a fumble.

At Stevenson Airport, aeronautical engineering boffins tested and adjusted operation of a new-fangled jet engine they had brought from experimental workshops in England, where Frank Whittle had invented the first such marvel. When the Canadian engineers saw that they had something very important and promising at hand they, of course, whisked it down to eastern Canada. And, in Toronto, this engine begat Chinook, the country's first original jet engine design; and Chinook begat Orenda 10, Orenda 10 begat Orenda 14, Orenda 14 begat Orenda Iroquois. And then, Orenda Iroquois

JIM SHILLIDAY | 155

begat oblivion from the Diefenbaker government in 1959, when the Arrow fighter bomber-defence project was abandoned because, they said, the manned bomber was obsolete, and missiles were the thing.

That first jet-engine scream in Winnipeg should have been Canada's klaxon call to the world that a new aviation-sheriff was coming to town; instead, it became a whine of defeat once again when Canadian entrepreneurial and technical skills within reach of the big-time were kneecapped by politicians.

Toronto's Paul B. Dilworth graduated in mechanical engineering from the University of Toronto in 1939, and joined the engine laboratory at the National Research Council, Ottawa. In 1942, his boss headed the team sent to the U.K. for aeronautical research. Their report included reference to a new gas-turbine engine for aircraft propulsion first tested in 1937 by Wing Commander Frank Whittle, of the Royal Air Force. His jet engine was the first to power an airplane in flight, May 15, 1941. A copy of the Whittle jet engine powered the first American jet, the Bell XP-59A, in 1941.

"At the same time, the RCAF was seeking ways to relieve Canada of dependence on the U.K. and U.S.A. for supply of engines for military aircraft produced in Canada," according to an article in the digital collection of *Library and Archives Canada*. This development in aero-engine technology opened up possibilities for Canada to become involved in the design and building of new power plants, and end dependence on foreign sources of supply. Today, that would mean the air force was "pro-active." As a result, it was decided to send a team to the U.K. to make an exhaustive survey of British jet engine development. Its mission was classified Top Secret.

The five-month survey involved the Canadian department of munitions and supply in England, the NRC Hydraulics Laboratories, and Paul Dilworth, who would have much to contribute. They investigated all jet engine research, development, and manufacturing in the U.K. The final report was issued in May, 1943, and laid the foundation for Canada's entry and rapid rise to become one of the world leaders in the development of gas turbine technology. And that's not said lightly.

The report recommended a cold weather test station in Canada to verify performance and compressibility limitations of jet engines at the low temperatures encountered in high-altitude flight. A high-priority meeting in June, 1943, included Fred Smye, of munitions and supply (soon to play a major role in the whole Avro-Orenda saga), and senior representatives from the RCAF and the research council. Dilworth helped set up the cold-weather test station at Winnipeg's Stevenson Field in just four months. The first engine test in Winnipeg was on January 4, 1944, after staff were trained in England. The Whittle-type Rolls Royce W2B ran without a hitch.

A Crown corporation, Turbo Research Ltd., was formed in March, 1944, in Toronto, to research, design, and develop gas turbine engines. While Dilworth stayed on in charge of the CWTS in Winnipeg, engineer Winnett Boyd and other engineers were transferred to Toronto from Ottawa, bringing advanced design work on a major plant for testing gas turbine compressor units. Apart from their short U.K. training, and at the CWTS, none of this team had prior experience in design or development of any kind of engines, reciprocal or jet.

Centrifugal and axial flow projects were tried out and, finally, the Chinook was built as a development and learning project. The Chinook and its successor, the TR5 Orenda, included a number of innovative design features that formed a solid foundation for later success. Said Paul Dilworth, in the *University of Toronto Magazine*: "Of possible archival interest, Winnett Boyd also designed the Orenda jet engine, successor to the Chinook. It became the leading high-performance military jet engine of the 1950s and early '60s. The Orenda saw service in the air forces of Canada, The Netherlands, West Germany, South Africa and Pakistan. After his short but brilliant career in jet engine design, Winn went on to design the outstanding NRU research nuclear reactor at Chalk River."

In the spring of 1946, the Canadian government transferred all work on gas turbines to private industry. In late 1945, A.V. Roe Canada Ltd. had been set up as an aircraft design and manufacturing facility in the former Victory Aircraft plant at Malton, near

Toronto. Its chairman was Sir Roy Dobson, and Fred Smye, Avro's first employee and the motivator behind all activities at Malton, arranged for Avro to take over and transfer the entire turbo operation to Malton. Paul Dilworth was appointed manager and chief engineer of the gas turbine division at Avro. The RCAF made it clear that it wanted a turbojet engine with a thrust equal to that of any contemporary engine in the world. By 1952, Avro had constructed a new plant, and after a year and a half 1,000 Orenda engines had been built. In total, 3,824 engines were delivered to the RCAF.

In 1955, A.V. Roe Canada Ltd. became the parent of two autonomous companies, Avro Aircraft Ltd. and Orenda Engines Ltd. Orenda went on to develop the famous Iroquois engine for Avro's supersonic Arrow. Both projects were considered to be beyond the state of the art at the time. The Iroquois was the most powerful jet engine in the world, rated at 19,250 lbs. of thrust, 25,000 lbs. with afterburner. It was at least 10 years ahead of its time, powerful enough to drive the ocean liner Queen Mary, and its technology is employed in most of today's high-performance engines of all types. After 7,000 hours of testing, to a simulated altitude of 70,000 feet, and a forward speed of Mach 2.3, the Orenda Iroquois program perished, along with the Arrow aircraft on Feb. 20, 1959 (almost the exact date of the fiftieth anniversary of the Silver Dart flight), as a result of the still controversial cancellation. Regrettably, Avro did not survive, but Team Orenda went on to provide other contributions to Canadian aerospace technology.

Wrote James H. Marsh, editor in chief of *The Canadian Encyclopedia*: "Diefenbaker's government rushed to buy American Bomarc missiles, which turned out to be useless without nuclear warheads, and then to purchase American

The first sound of a jet engine in Canada shrieked outside a hangar similar to one of these at Winnipeg's Stevenson Field, in 1944. *Photo: Courtesy Brian Johannesson*

JIM SHILLIDAY | 157

CHAMBERLIN'S GENIUS: CANADA REJECTS, U.S. EMBRACES

MICHAEL COLLINS, U.S. COMMAND module pilot for the Apollo 11 moon landing, wrote in Liftoff (1988), that NASA was lucky Canada cancelled the CF-105 Arrow in 1959 because it provided them with 25 highly-skilled and badly-needed aeronautical engineers.

Jim Floyd, former Avro chief engineer and one of the developers of the CF-105 Arrow, wrote that NASA's Space Task Group at the Langley, Virginia, facility, led by Robert Gilruth, former assistant director of the U.S. National Advisory Committee for Aeronautics, was overloaded with urgent work on the Mercury space capsule design. Before the Arrow's cancellation, Gilruth had helped test Arrow models at Langley, and borrowed the unique Canadian team to work on the Mercury project.

His idea was to keep the ex-Avro engineers together as a team, with the intention of returning them when Avro sorted out the future of the company after the Arrow cancellation. But when Avro Canada folded, the team remained with NASA, now integrated in the space program, and later contributing to both the Gemini and Apollo projects. One of the American space contingent praised the Canadian engineers: "They had it all over us in some areas ... just brilliant guys ... they were more mature and were bright as hell and talented and professional to a man."

Arrow engineer Owen Maynard, born in Sarnia, went to NASA and designed the Apollo 11 landing module for man's 1969 first moon landing, and then return to Earth. The module's legs were designed and made in Quebec.

It is absolutely astounding what a difference these Canadian aeronautical experts made to the whole U.S. space program. In Avroland, Copyright AvroArrow.Org, a digital collection, an article by Jim Floyd titled Canada's Gift To NASA: The Maple Leaf in Orbit, states that Canadian Jim Chamberlin, one of the Avro engineers, was honoured in the 1960s for his contribution to the Gemini program. The certificate read:

"National Aeronautics and Space Administration (NASA) Manned Spacecraft Centre presents this certificate of commendation to James A. Chamberlin for his outstanding contribution to this nation's space flight programs, for the technical direction and leadership of the Project Mercury, for his creation and promotion of the Gemini concept and for his guidance in the design of all manned spacecraft used in the United States exploration of space to date." Signed: Robert R. Gilruth, Director, Manned Spacecraft Centre, Houston, Texas.

This great honour was bestowed on a man the Canadian government threw away. Aerodynamicist Jim Chamberlin, from Kamloops, B.C., was a graduate of the University of Toronto and was design chief at Avro, where he oversaw development of North America's first jet transport, the C-102 Jetliner (also thrown away), the CF-100 (Canuck) all-weather fighter, and the Arrow (yes, thrown away).

James A. Chamerblin (right), with fellow Avro engineers.

Crowds gather in awe as CF-105 Arrow is rolled out on Oct. 4, 1957, at the A.V. Roe Canada plant, in Toronto.

Chamberlin was appointed leader of the Canadians at Langley, and became Gilruth's main advisor. He had a huge role in the design of the Mercury capsule that put John Glenn into orbit on Feb. 20, 1962—three years after the Arrow's cancellation. He soon rose to head of engineering and administration on Mercury, then head of the U.S. Space Task Group's engineering division. In this role, he directed multi-million-dollar projects such as the two-man Gemini project and the Apollo moon lander. Neil Armstrong's dramatic words on first stepping onto the moon's surface: 'One small step for man, one giant leap for mankind,' owed one huge debt of gratitude to Chamberlin, wrote Floyd. Chamberlin received the NASA Gold Medal for his work on Gemini and was described by a NASA administrator as one of the most brilliant men to work with NASA. At the time of his death in 1981, he was technical director for McDonnell at the Johnson Space Centre, in Houston.

"On a wintry day in February, 1962, as four million Americans watched with pride as a cavalcade of limousines slowly made their way along the streets of New York to wild cheers and mountains of ticker tape, America's first astronaut to orbit the earth, John Glenn, waved in appreciation to the crowds from the first car. In the second car was a quiet Canadian from Kamloops, James A. Chamberlin. Oh Canada!!"

A LEADER IN SPACE ROBOTICS

AVRO'S PAUL DILWORTH LATER founded a major design and development company for aviation wind tunnels and automotive environmental test facilities for international clientele. He was also instrumental in bringing to Canada the NASA Space Shuttle manipulating arm project, and played an important role in its development as the famous Spar Aerospace "Canadarm" used with outstanding success on all NASA Shuttle flights.

"The virtually faultless operation of the arm," wrote Jim Floyd, "has been a major influence in the successful launch of a large number of important projects over fifteen years of Shuttle operations." Canada's aerospace companies have been so impressive, the European Space Agency asked them to build the Mars Rover for its Mars mission. But look out! Although the project would cost just $100 million over ten years, and need only re-allocation of funds already set aside for the Canadian Space Agency, the federal government indicated in 2008 that the Canadian participation would not happen—and everything hit the fan.

One aeronautical engineer protested that Canadian space companies, some of the leading robotics companies in the world, were threatening to leave the country and head to the United States. The loss of these companies, he warned, would perhaps be viewed in a few generations as a watershed event, much in the way the Diefenbaker government's decision to end the Avro Arrow project is viewed today.

Fortunately, the politicians saw the light. Canada is still in the European Mars mission, and in May, 2008, a NASA spacecraft landed on the Red Planet's northern polar region after a 10-month voyage. To help in a successful search for water on Mars, a $37-million weather station a little bigger than a shoebox, and displaying a small maple leaf flag, was the first Canadian scientific instrument to make it to an alien world. A Canadian scientific team would study the data sent back from Mars.

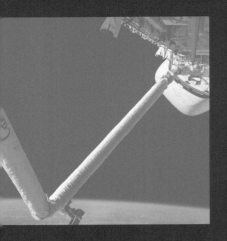

Spar Aerospace "Canadarm".

Voodoo interceptors that the air staff had judged inferior years before. For many Canadians, the cancellation of the Arrow was a mortal blow to part of the national dream and confirmation that our leaders did not have the courage or the vision to forge a coherent defence policy independent of the United States."

In 1959, Canada selected the Lockheed F-104 Starfighter as its new dayfighter, built under licence by Canadair, with Orenda making the engines. In 1962, it began building engines for the Canadair CL-41 Tutor (Snowbirds). In 1967, Canadair started putting together Northrop F-5s, with Orenda supplying the engines. Great work for Canadian firms, but none of the designs—airframes or engines—were Canadian, except the CL-41. Which raises the question: if civil and military aviation had been encouraged in earlier years, might all of the aircraft and engines—now being supplied by private companies (after cancelling the Arrow, Canada funded civil aerospace instead), such as Bombardier and Allied Wings, in a co-operative business deal with the Canadian air force costing many billions of dollars—have been designed and built by Canadian companies?

What role has politics played in this equation? Is it a good deal for taxpayers?

Steeply rising costs, and predictions in a 1957 British defence White Paper, of the imminent demise of the manned fighter, may have been an impetus to the 1959 scrapping of the Avro Arrow project, with little apparent recognition of the fact the Arrow's death—a design many years ahead of its time—meant a crippling blow to future aeronautical development in Canada, and loss of a 30,000-person expertise, with all their technological advances, many eagerly scooped up by the fledgling American NASA program established just a year earlier, and U.S. aircraft builders. It meant that Canada's short time as a ranking military power was over. That year, Canada was enthusiastically observing its fiftieth anniversary of powered flight.

BOMBARDIER: TAKING ON THE WORLD

THE FOURTH-LARGEST aviation/aerospace sector in the world is here in Canada. The Canadian Aviation Maintenance Council points out that Canadian firms are global market leaders in regional aircraft, business jets, commercial helicopters, small gas turbine engines, flight simulation, landing gear and space applications. About 137,000 persons work in the industry.

Canada's civil aircraft industry is the sixth largest in the world. Exports totalled $5.2 billion, representing approximately 75 percent of sales. Total direct employment was approximately 41,000 persons. In 2008, the Conference Board of Canada predicted the strongest profit levels in years for the Canadian aerospace manufacturing industry. Dominated by Bombardier, this sector had a 2007 revenue of $9.7 billion out of a total for the country of $14.9 billion. Airlines around the world were retiring older, less fuel-efficient aircraft, and many were expected to invest in Bombardier's fuel-efficient Q400 NexGen turboprops.

For decades, an aviation phenomenon has been building in Canada, well, more precisely, in Quebec—Bombardier Aerospace. The average Canadian is not sure how a company manufacturing Skidoo snow machines has become one of the word's aviation leaders. It appears that, from the early days of total lack of interest in civil aviation, the pendulum has swung until politicians and entrepreneurs jostle to wield power and influence to make companies such as Bombardier a success. There have been whispers for years that politicians in Ottawa have been too helpful to Quebec's aero industry, particularly Bombardier, in matters of loans and grants and whatever else.

Some complain that political wooing of Quebec has harmed the country as a whole. They suggest that meddling politicians have corrupted the natural development of western Canada as a centre for aviation expertise. The country's main airlines were born in the west, but when the aerospace industry there was ready to expand, Ottawa stepped in. Trans-Canada Airlines (Air Canada's ancestor) was headquartered in Winnipeg. In the 1950s, its maintenance section was transferred to Montreal amidst great outcry; little by little, the company's head office operations were moved to that city. The politicians gave to Montreal what was traditionally Winnipeg's.

Winnipeg's sense of betrayal was heightened in 1986 when Brian Mulroney's Conservative government handed the CF-18 overhaul contract to Quebec. In his 1992 report, Assistant Auditor General David Rattray pointed out: "In 1986, a consortium including Bristol Aerospace, of Winnipeg, was found to have the best over-all bid for the systems support and airframe repairs and overhaul contract for the CF-18. As events unfolded, however, the contract was awarded to Canadair (a Crown corporation the federal government later handed over to Bombardier), the leader of the consortium that had ranked second...." This egregious political manipulation once again cost Winnipeg a traditional industry, in favour of Montreal.

Jeffrey Simpson, columnist for the *Globe and Mail*, wrote: "The awarding of the CF-18 maintenance contract provided a classic illustration of the inflationary consequences of political bribery...Despite experts' clear verdict, the prime minister and other Quebec ministers steered the contract to Canadair, devising a rationale about technology transfer to camouflage a purely political decision."

Maclean's columnist Andrew Coyne commented, wryly, "Oh sure,

Coyne commented, wryly, "Oh sure, it makes passenger aircraft, which it sells around the world. But its *business* is collecting subsidies."

THE EVOLUTION OF BOMBARDIER

THE FIRM PURCHASED Canadair Ltd. (which started in the 1920s as the aircraft division of Canadian Vickers Ltd., in east Montréal), from the Mulroney government in 1986. Canadair was then the biggest civil aircraft manufacturer in Canada, having built 4,400 civil and military aircraft (e.g. 1,815 F-86 Sabres; and much earlier, a series of airplanes, starting with the Vedette, the first Canadian flying-boat design; and Canso amphibians in wartime), and it became the core of Bombardier Aerospace.

Canadair had been an important producer of executive jet aircraft such as the Challenger series and the Regional Jet, which pioneered the

Dash 8 in flight over Montreal.

commuter jet market. In 1989, Bombardier absorbed Short Brothers (Sunderland flying boat), of Belfast, Northern Ireland, the oldest established plane builder anywhere. In 1990, the pioneer business jet builder, Learjet, of Wichita, Kansas, was acquired. Bombardier picked up 51 per cent of fabled de Havilland Canada (Twin Otter, Dash 7 and 8 turboprops) from Boeing in 1992, became sole owner five years later. The Chipmunk trainer was de Havilland's first design, then the Beaver, in 1947, which set the trend for several short-takeoff-and-landing (STOL) aircraft: the Otter, Caribou, Buffalo, Twin Otter and Dash 7. The Dash 8, a market leader in turboprop commuter aircraft, followed.

In a move to end a dual trans-Atlantic monopoly, Bombardier took on world aerospace giants Boeing and Airbus at the Farnborough Air Show in July, 2008, announcing its CSeries, 110-130 seat passenger jet would be partly-made and all assembled at Mirabel, near Montreal (beating out Kansas City), and enter service in 2013. Coming at a time when rocketing fuel prices and a shaky economy had put the brakes on a boom in airline orders, the CSeries was reputed ready to "revolutionize" the economics of its smaller-size range due to fuel-efficient, green technology.

The CSeries, its largest aircraft to date, would sell for $46.7 million each. Though no contract had been signed at the time, longstanding Bombardier client Lufthansa had provisionally ordered 30 planes with an option for 30 more. Potential customers also included China Southern, Shanghai Airlines, ILFC (International Lease Finance Corp.) and Qatar Airways. The Bombardier project was supported by $350 million of federal funding, and more than $100 million from Quebec. Both governments would get a return on each aircraft sold and, to the joy of politicians, bragging rights for about 3,000 new jobs. Bombardier's worldwide operation brought in additional funding from Northern Ireland (design and partial manufacturing of CSeries) and Britain.

Industry Canada says the government has seen the wisdom of investing in aerospace companies, as do governments around the globe. "One of the primary drivers for government intervention is the link between the industry and national security. The economic characteristics of the industry with its high development costs, high risks and long payback periods, combined with the highly cyclical nature of the sector, have also been driving forces calling for government involvement. These characteristics make it difficult for the private sector to shoulder all financial requirements alone and require the government to intervene to support the sector.

Predictably, the U.S. applied to the World Trade Organization for ruling on whether or not the financial support for Bombardier was illegal under international law, as it insists are billions of euros given to support Airbus. The European Union claims the U.S. subsidizes Boeing with military and space program budgets.

Six years ago, the WTO supported a U.S. claim that a $1.7 billion low-interest loan by Canadian taxpayers to Bombardier broke international trade rules.

The Aerospace Industries Association of Canada, representing more than 400 aerospace companies, cautions that intensifying global competition and constantly-changing market dynamics pose risks to Canada's established market positions and threaten their ability to seize new market opportunities. Said an industry association spokesman: "With extraordinary support from their respective national governments, competitors in the United States, Europe, South America, Japan and Asia are aggressively challenging Canadian firms in key market niches and at every level of the supply chain. As the new century unfolds, profound changes are sweeping the industry. Other nations, notably the United States and European countries, continue to make sizeable investments to transform their aerospace industries to meet the competitive challenges of the next decade and beyond."

That's a reasonable prod, but perhaps still a little myopic if increased government investment in Canada's aerospace industry continues to ignore other parts of the country. What a fine thing it would be if the Big Thinkers in the industry, in the federal cabinet, and in the civil service, would concentrate on the idea that there can be too many eggs in one basket, and that there is a lot more to Canada than just Ontario and Quebec. It seems pretty likely that a plane with two engines performs better when one engine isn't allowed to run out of juice. But Canadians pride themselves on being fair-minded. Maybe what has gone around will come back around again?

WESTERN CANADA'S IMPORTANCE TO COMMERCIAL AVIATION

WARDAIR WAS A WESTERN CANADIAN bush plane operation started by Max Ward that became a big player in North America. It flew domestically in B.C., Alberta, Manitoba, Ontario and Quebec, also developing international routes to Europe, U.S.A., Caribbean and South America. From biplanes flying out of Edmonton and Yellowknife in the 1950s, Wardair grew to use giant A310 Airbuses by 1989, when it was sold to Canadian Airlines International of Calgary. In 2001, CAI was acquired by Air Canada. Western Canada's latest contribution to commercial aviation is WestJet, also of Calgary, a low-cost airline with routes into the U.S., Mexico and the Caribbean. Second in size to Air Canada, it aims to be one of the world's top five in profitability by 2016.

A DIVERSIFIED INDUSTRY

THE SECOND LARGEST aerospace company in the country (after Bombardier) is Pratt & Whitney Canada Inc. Since it developed the PT6 turboprop engine in the 1960s, its jet turbines account for more than 50 percent of small and medium-sized engines manufactured in the world. Set up in 1929 to overhaul Wasp radial engines, it grew during the Second World War through sales to the military.

With annual revenues upwards of $3 billion, western Canada's aerospace industry is also showing renewed promise, according to Western Economic Diversification Canada. An estimated 17,000 persons are directly employed in the industry, the majority of them contributing engineering, technical or management expertise. Western Canadian firms such as Standard Aero, Boeing Canada and Bristol Aerospace (Magellan) specialize in repair and manufacturing of regional and commercial aircraft; maintenance, repair and overhaul of small gas turbine engines; and provision of parts and service for flight simulators. Accounting for 14 per cent of Canada's total $22 billion industry performance, they are a key reason why our industry is the fourth largest in the world, behind only the United States, Britain and France.

LOOKING AHEAD

BRITISH AUTHOR ANTHONY Burgess wrote that John Kenneth Galbraith and Marshall McLuhan were the two greatest modern Canadians that the United States has produced. Tongue in cheek, but he could have included NASA genius Jim Chamberlin and colleagues. Is it time that Canada developed an environment capable of keeping its great minds at home? If the twentieth century didn't belong to Canadian aviation, to paraphrase Sir Wilfred Laurier, the twenty-first Century could—particularly if we can alter our mind-set. The continued prospering of this country will depend on those who have "ridden the wings of our people's cunning." •

 Piloted by Vintage Wings's Tim Leslie, gleaming-gold Hawk One (now a 55-year-old Cold War retiree) tucks into famed Snowbirds' formation while practising for the centennial show season at Comox, B.C., in February, 2009. *Photo: Derek Heyes*

POST-LANDING CHECK

VINTAGE WINGS AND
HAWK ONE

THOSE WHO FLY KNOW THEY don't have to die to go to heaven. They don't even have to be good. The world above is not the realm of man, he is just a visitor there. Awestruck, he is reluctant to leave, though he knows he must.

Rising from the runway one day in 2009, a Sabre lifted its nose sharply as though sniffing the air, then leapt upwards, climbing to the rarified altitudes where oxygen and fuel combined, exploded and propelled at three times the power they could produce at sea level. Beneath the dark dome of high altitude, the jet fighter's flight path traversed snowy peaks; prairie vastness; grey blotches of big cities; the waters of myriad small and great lakes glinting—a vast tableau known as Canada.

The jet fighter up there was Hawk One. The Canadair Sabre had been the first Canadian-made jet fighter, flown by Canadian pilots, to engage in close live combat with other fighter jets, a record that stands to this day. The glory of its swallow-like flight in combat conditions was polished to gold-like brightness during Canada's centennial of flight year, when the Hawk One reincarnation of the Sabre soared and flitted the heights of this country. Its magnificence shone for hundreds who have such memories of sky, and for those thousands who simply were in the thrall of beautiful flying machines and the mystique of flight. And they owed the moment, in some degree, to two men bitten by the aviation bug.

IN PURSUIT OF HERITAGE

LIKE WEALTHY ENTREPRENEUR Alexander Graham Bell before him, Michael Potter saw a challenge that had wings, and he took off in pursuit. Where Bell helped invent new flying creations to kick-start a continental aviation industry, Potter began buying and restoring to flying status past glories that had fallen on hard times, with the goal of showing Canadians the wonders of their aviation heritage. One aircraft in particular—an early Cold War Canadian Sabre fighter—he rescued and renamed Hawk One and made it the "poster

JIM SHILLIDAY | 165

plane" of Canada's 2009 centennial of flight celebrations.

Potter, computer and software magnate cum aviator and "living air museum" founder, joined wingtips with the Canadian air force and the aerospace industry to give his country's centenary air shows a sonic lift. Aviation enthusiasts all over Canada were treated to special flying presentations through 2009 as the country commemorated its century in the air, along with the memory of its early Cold Warriors in their Canadair Sabres, and the country's advanced position in aerospace.

Founder of software giant Cognos, the country's largest such firm, and enjoying the largesse of success, Potter decided to do Good. Born in London, England, in 1944, he fell in love with airplanes after flying gliders at Royal Military College and UBC in the 1960s. He decided to become an aviator for business travel—a feat he accomplished far beyond the usual boundaries. He has an air transport licence on which he has accumulated an impressive 4,500 hours flying time. He now owns 18 flyable classic aircraft, most of them warplanes—just one a jet—has checked out in most of them and has become a major Canadian showman. In fact, he hit the silver screen in the 2009 film starring Hilary Swank about the life of Amelia Earhart who took off from Newfoundland and was the first woman to fly the Atlantic. In period costume, he flew his vintage WACO Taperwing biplane in film sequences.

Back in the mid-1990s, Potter decided to retire at age 51 from Cognos and pursue his passion for aviation. In 1998 he bought his first heritage plane, a Beech Staggerwing, soon followed this up with other aircraft. When he purchased his favourite, a wartime Supermarine Spitfire (repainted in the wartime aircraft markings of 421 Squadron), "I asked the Canada Aviation Museum in Ottawa

Mike Potter in his beloved Chase 2 Spitfire thunders down the left side of Runway 27 still with an eye on Hawk One's landing at Gatineau airfield by Paul Kissman, after the reconstituted Cold War jet fighter had completed its first solo. *Photo: Peter Handley/Vintage Wings*

VINTAGE WINGS STABLE OF STEEDS

VINTAGE WINGS STABLE TODAY: Supermarine Spitfire XVI (Potter's flying favourite, in almost mint condition); a second Spitfire; de Havilland Tiger Moth and Chipmunk; Beech Staggerwing; North American P-51 Mk IV Mustang (must be flown cautiously: a wing can stall without warning at almost any speed or attitude); Hawker Hurricane Mk IV (also tricky to fly); a second Hurricane, Mark XII; WACO Taperwing A.T.O., with original powerful Wright engine (a "light-footed" plane, favorite of 1920s aerobatic pilots); de Havilland Canada DHC-2 Beaver; North American Harvard; Fairey Swordfish Mk III; Curtiss P-40 Kittyhawk, former RAAF; de Havilland Fox Moth, originally owned by Edward, Prince of Wales; Goodyear FG-I D Corsair painted in Fleet Air Arm colours to honour Robert Hampton Gray, winner of war's last VC in 1945 while attacking Japanese shipping; Hawker Hurricane Mk XII; Bellanca Citabria; Westland Lysander expected to fly in 2009; Canadair Sabre 5.

to provide their facility for an unveiling of the 'Canadian Spitfire.'

To my amazement, two or three thousand people came—the parking lot overflowed, there was standing room only in the museum. Many air combat veterans were seated in the front rows for the short ceremony. Then, after a flying display that brought the unique growl of a Merlin powered Spit back to the former site of historic RCAF Station Rockcliffe after decades of absence—many of those veterans climbed onto the wing and settled into the cockpit of Spitfire SL721. Memories were revived, tears flowed, families looked upon grandfathers with renewed admiration, and thoughts turned to the brave young men from Canada who flew in combat, many of whom lost their lives in the fight."

Potter decided to gather all his planes under one roof and create the Vintage Wings of Canada Foundation, in French, La Fondation des Ailes d'Époque du Canada. Potter estimated that his aviation foundation absorbed about a third of his time, the rest being taken up with investing in family, finances, charities and business work. The not-for-profit foundation is devoted to preserving and operating the largest collection of privately-owned, airworthy vintage wartime aircraft in Canada. Vintage Wings is not a museum; it could be called a "live museum", an airborne showcase of beautiful flying memories—the past in the present.

The Vintage Wings collection of classic aircraft is housed in a new $3 million 23,000-square-feet hangar at Ottawa/Gatineau Airport, of traditional 1930s-style design, with complete maintenance facilities, aviation library, workshops, offices and amenity spaces. It is not open to the general public, but there are occasional Open Hangar days, group tours can be pre-arranged, and visits from aircraft and veterans are welcomed. Flying demonstrations include Battle of Britain events every September, and Classic Air "Rallyes".

Michael Potter is unabashedly proud of his country, its accomplishments in aviation, and is doing what he can to make Canadians, young and old, aware of their aviation heritage and their air force veterans. He believes "that both the new and old generations will always have an interest in the history of aviation and that these combat planes will remain mythical machines with a story to tell about the role our country played in insuring victory and freedom." The foundation's three main purposes are "to commemorate the achievements of our veterans, educate Canadians and inspire our youth."

In 2006, the Canada Aviation Museum announced a collaboration with Vintage Wings which would

JIM SHILLIDAY

 The Hawk One pilots—and guest: Tim Leslie, test pilot Paul Kissmann, Fern Villeneuve (1960s Golden Hawks leader), Dan Dempsey, Steve Will, Canadian astronaut Chris Hadfield. *Photo: WO Serge Peters, CAF*

have Potter's aircraft, some of the rarest and most historically significant in the world, putting on flypasts for Canada Aviation Museum visitors every Friday afternoon. Said Potter at the time: "We believe that showcasing these marvellous aircraft is a most effective way to gain the attention of young people and direct their interest into science, engineering, military and political history and give them an appreciation of the lives of many outstanding Canadians who have shaped our country in a most extraordinary way."

'SPITFIRE OF THE JET AGE'

WINGED MARVELS CRISS-CROSSED the multi-thousand kilometer width of Canada in 2009 to celebrate the century of flying. The early Cold War Sabre jet fighter had been snatched from the scrap heap and reconstituted by volunteer technicians as the hybrid and gleaming Hawk One. This Mark 5 Canadair Sabre—known as the "Spitfire of the Jet Age", and one of the last "real" fighter planes—was a veteran of NATO service in Europe, and served with the famed Golden Hawks aerobatic team. After 16 years military service, it was sold to the private sector and half a dozen different American owners. It mouldered another 10 years in mothballing and then—resurrection.

Some elegant old warplanes, a couple of astronauts and a replica of the Silver Dart—a bamboo-and-fabric "aerodrome" that was the first airplane to fly in the British Empire—co-starred in Canada's anniversary celebrations. Chris Hadfield, former CF-18 pilot and Canadian Space Agency astronaut, joined four other highly experienced former Canadian Forces pilots to fly the F-86 Sabre at air shows across Canada throughout 2009.

The Hawk One project was a spin-off of Vintage Wings of Canada, set up in collaboration with the Department of National Defence and mostly supported by corporate sponsorships (such as Westjet, Marks Work Wearhouse, Cirrus Research Associates, Magellan Aerospace and Inter Pipeline) but no government grants. The jet was loaned for the 2009 showtime. It flew at selected air shows alongside Blue Hornet, an air force CF-18 front-line fighter—painted in high gloss red, white, blue and gold— and Canada's present jet aerobatic team formed in 1971, the Snowbirds. Their CL-41 Tutor jets put on 100 flying shows during the 2009 centennial season. Most of Potter's other warplanes took part in centennial airshows within flying range of Ottawa.

Crowds cheered at Hawk One's roll-out ceremony on Sept.20, 2008, stirred by Spitfire, Hurricane, Mustang and Corsair flypasts, the Canadian Warplane Museum's Lancaster bomber (one of only two in the world that fly), and a great formation show by the Canadian Harvard Aircraft Association. Fern Villeneuve, first commanding officer of the Golden Hawks, was there. When a test flying period was finished, the Sabre flew to No.4 Wing, Canadian Forces Base Cold Lake, Alberta, to be painted in the bright gold with red trim of

Former Canadian astronaut Bjarni Tryggvason gets the Silver Dart off the ice during flight tests on Feb. 22, 2009, in an historic re-enactment at Baddeck, N.S.. *Photo: Janet Trost, Hawk One*

SILVER DART REPLICA FLIES

A GROUP OF ENTHUSIASTS calling itself the Aerial Experiment Association 2005, began building an exact-scale replica of the Silver Dart in Welland, Ontario, sponsored by two aviation companies: Pratt and Whitney Canada and Leavens Aviation Inc., as well as the Ontario Trillium Foundation. Fittingly, one of the volunteer workers was Gerald Haddon, of Oakville, Ont., Douglas McCurdy's grandson, and the only person alive today who knew Canada's First Pilot intimately. He says McCurdy was more of an influence on his boyhood development than was his father. In the salt water of Bras D'Or Lake, young Haddon learned from his grandfather the skill of sailing a boat, along the same stretch above which Silver Dart made its historic flight. Haddon is a man dedicated to, and invigorated by, the memory of his boyhood hero, Douglas McCurdy.

The replica was built to the same standards as the original, with nearly the same materials. The volunteers worked hard for four years. Just a couple of weeks before the anniversary, the Silver Dart was finished and rolled out at the Hamilton Airport, to suitable fanfare. Shortly, it was successfully test flown, with McCurdy's grandon describing the action for CTV from a helicopter above. Then it was pulled back indoors, partially disassembled and readied to be shipped by truck to Sydney, Nova Scotia (where, the following July, McCurdy's grandson made the key speech at the renaming of the Sidney airport, the "J. A. Douglas McCurdy Sydney Airport".

Thousands were on hand at Baddeck for the historic re-enactment on February 22, 2009, one day ahead of the 100th year because of bad weather. Former astronaut Bjarni Tryggvason piloted the replica that will end up either in the Bell Museum in Baddeck, or the Toronto Aerospace Museum. More than 30 museums across Canada are solely dedicated to flight. Winnipeg's Western Canada Aviation Museum has the last of 1,815 Sabres built; a Mark Six Sabre, that saw action with the Pakistan air force. Replicas of the Silver Dart are housed in the Atlantic Canada Aviation Museum, Halifax; the Canadian Bushplane Heritage Centre, Sault Ste. Marie; the Aerospace Museum, Calgary; the Reynolds Alberta Museum, Wetaskiwin, and the Canadian Aviation Museum, Ottawa.

Newly painted for centennial shows, air force CF-18 Blue Hornet emerges from the paint bay at CFB Cold Lake, Alberta, in February, 2009, in a design by Jim Belliveau. *CF photo*

the RCAF Golden Hawks aerobatic team that lit up five air-show seasons beginning in 1959 to mark the 50th anniversary of flight.

Hawk One's first test flight was in early November, 2008, piloted by Paul Kissmann. The vast network that Vintage Wings can draw on paid off when Transport Canada regulations ruled that a Sabre-experienced pilot had to be in attendance and on the radio supervising. Retired Brig.-Gen. Paul A. Hayes, a member of the Sabre Pilots Association of Air Division Pilots (SPAADS), that has hundreds of members, got an urgent request from Kissman saying they were ready to fly, but none of the planned pilots had flown the Sabre. Hayes had already provided them with Sabre 6 pilot operating instructions and the checklist. He had talked to them about the Sabre's operation and performance. So he was the man they needed. He flew himself to Gatineau from Toronto, attended the flight briefing, monitored the walk-around and start, and was at the debriefing. The SPAADS membership list shows Hayes's rank as Air Commodore. The Canadian forces were unified in 1968 and assumed army ranks, but SPAADS prefers to show the former RCAF (RAF) rank structure.

Five pilots flew the Sabre during the 2009 celebration appearances. Lt. Col. Steve Will—former leader of the Snowbirds aerobatic team and a CF-18 squadron commander, who oversaw the Hawk One project—along with former Chief Fighter Test Pilot for Canada's Air Force Paul Kissmann, now a research test pilot at National Research Council in Ottawa, provided the aerobatic demonstrations. Pilots Dan Dempsey (another former Snowbirds lead pilot) and Tim Leslie, ferried the aircraft from location to location. Leslie, who originated the Hawk One idea, is a former military pilot, now a test pilot with the National Research Council; he is also vice-president and chief of operations at Vintage Wings. Vintage Wings founder Michael Potter flew selected missions.

Hawk One's 2009 flying schedule included appearances at Abbotsford, British Columbia; Moose Jaw, Saskatchewan; Toronto, Ontario; Moncton, New Brunswick; and the capital, Ottawa (July 1st, Canada Day). Colonel (Ret.) Chris Hadfield did selected flypasts, such as Canada Day in Ottawa, and the extra-special February anniversary flight over Baddeck, Nova Scotia, where the first Canadian flight took place in 1909.

The Canadian air force established a Centennial of Flight project office to co-ordinate its own events and activities and to facilitate communication with many organizations across the country planning to observe the anniversary.

To mark the centennial, more than 125 private aircraft (some homemade, some more than half a century old) registered for a Century Flight in July from airports around Vancouver, B.C. to Sydney, N.S., close to Baddeck. They hop-scotched across the nation, stopping in Calgary, Brandon, Marathon, Brampton, Sherbrooke and Fredericton. Lifting off from Boundary Bay was a 1944 Noorduyn Norseman, the first Canadian-designed and built bush plane. Flown by two brothers (Dennis and Greg Mockford, of Alberta, both seniors), it stopped off in Red Lake, hallowed centre of Canadian bush flying—a dozen Norseman's still fly there—to attend a festival dedicated to the Norseman. Really caught up in centennial spirit, the Mockfords already had flown their plane on a 16,000-kilometre circumnavigaton of Canada starting in Red Lake in May.

Rebuilt Hawk One, a 1950s-era Sabre with Paul Kissmann (National Research Council's chief test pilot) at the controls, lifts off from Canadian soil for the first time at Gatineau. It then was given the Golden Hawks paint job. *Photo: Peter Handley/Vintage Wings*

HAWK ONE WAS 55 YEARS OLD

THE F-86 THAT WAS THE backbone of the Centennial Heritage Flight was a Canadair Sabre 5 manufactured in 1954, retrofitted by Vintage Wings with wings equipped with leading edge slats and an Orenda 14 engine, upgrading it to Sabre 6 class. The original wings, damaged in an accident, were replaced with those from a U.S. F-86F.

Canadian Sabre squadron pilots flew CL-13 (Canadair's designation) Marks Two, Five and Six. The Sabre Mk 2 closely matched those flown by Americans, having the same engine pushing out 5,200 pounds of thrust.

But it was the Canadian designed and built Orenda engines that made the Canadian Sabre so superior to the American version. The Orenda 10 in the Mk 5 provided 6,500 pounds of thrust. This output soared to 7,300 with the Orenda 14 in the Mk 6. Ultimately, Canadair built six variants of the Sabre, the most famous and capable being the Sabre 6. The aircraft had a top speed of 710 mph and a service ceiling of 55,000 feet. The first Sabre 6 took flight on Oct. 19, 1954. The last F-86 Sabre rolled off the assembly line at Canadair on Oct. 9, 1958.

A CALL TO ACTION

VINTAGE WINGS OF CANADA Foundation heard the call to action and has scrambled to meet the foe—decades of low official interest in promoting Canada's aviation potential and heritage. Wayne Ralph encapsulated the challenge in his fine biography, *William Barker VC*: "… Canada's history as a Dominion of the British Empire is no longer taught in Canadian schools, and our history as warriors within that empire is now politically and socially irrelevant and, in some eyes, horrific and contemptible." Vintage Wings is gung-ho and well equipped to help win the battle. No bird soars too high, said William Blake, if it soars with its own wings. That's what Canadian aviation needs to do. •

Vintage Wings test pilot Paul Kissmann before first flight of Hawk One. *Photo: Peter Handley/ Vintage Wings*

Reference Books

Reluctant Genius: The Passionate Life and Inventive Mind of Alexander Graham Bell, by Charlotte Gray, HarperCollins Publishers Ltd., Toronto, 2006.

The Silver Dart, by H. Gordon Green, Atlantic Advocate, Fredericton, 1959.

Canadian Aviation Since 1909, by K.M. Molson and H.A. Taylor, Canada's Wings, 1982.

The Creation of a National Air Force: The Official History of the Royal Canadian Air Force, Vol. II, by W.A.B. Douglas, U of T Press, Toronto, 1980.

Barker, VC, by Wayne Ralph, Doubleday, Toronto, 1997.

The Royal Flying Corps in World War I, by Ralph Barker, Robinson, London, 1995.

West With The Night, by Beryl Markham, North Point Press, San Francisco, 1987.

Pure Luck: The Authorized Biography of Sir Thomas Sopwith, 1888-1989, by Alan Bramson, Patrick Stevens Ltd., 1990.

Canada's Air Force Today, Larry Milberry, CANAV Books, Toronto, 1987.

AIRCOM, Canada's Air Force, by Larry Milberry, CANAV Books, Toronto, 1991.

There Shall Be Wings: A History of the Royal Canadian Air Force, by Leslie Roberts, Clarke, Irwin, Toronto, 1959.

A Thousand Shall Fall, by Murray Peden, Canada's Wings, Stittsville, 1979.

Rites of Spring: The Great War and the Birth of the Modern Age, by Modris Eksteins, Lester & Orpen Dennys, Toronto, 1989.

Canada's Flying Heritage, by Frank H. Ellis, University of Toronto, 1961.

A Military History of Canada: From Champlain to the Gulf War, by Desmond Morton, McClelland and Stewart, Toronto, 1992.

Customs and Traditions of the Canadian Armed Forces, by E. C. Russell, Deneau & Greenberg, 1980.

Straight On Till Morning, by Beryl Markham, St. Martin's Press, New York, 1987.

The Plan, Memories of the BCATP, by James N. Williams, Canada's Wings, Stittsville, 1984.

High Flight: Aviation and the Canadian Imagination, by Jonathan F. Vance, Penguin Canada, 2002.

Fighting Airman, by Major Charles J. Biddle, Doubleday, New York, 1968.

1001 Flying Facts & Firsts, by Joe Christy, Tab Books, 1989.

Whose War Is It? How Canada Can Survive in the Post-9/11 World, J.L. Granatstein, 2007

242 Squadron, The Canadian Years: being the story of the RAF's all Canadian Fighter Squadron, by Hugh Halliday, Canada's Wings Inc., 1981.

Pioneers of Flight, by Henry T. Wallhauser, Hammond, 1969.

V.C.s of the Air, by John Frayn Turner, Clarke, Irwin and Co., Toronto, 1960.

The Soviet Air and Rocket Forces, edited by Asher Lee, Frederick A. Praeger, New York, 1959.

A History of Soviet Air Power, by Robert A. Kilmarx, Frederick A. Praeger, New York, 1962.

Van Sickle's Modern Airmanship, edited by John F. Welch, TAB Books Inc., 1981.

Soaring wings: A biography of Amelia Earhart, by George Palmer Putnam, Harcourt, Brace and Company, 1939.

Publications & Web Sources

Re-connecting Canada to the World – via Europe? Dr. Julian Lindley-French in International Journal Vol 59. No, 5 (CIIA: Toronto), Haglund, David (Ed.), 2005.

Airforce Magazine, Air Force Association of Canada.

CAHS, The journal of the Canadian Aviation Historical Society, Vol.4, No.4, 2006.

No. 1 Fighter Wing's magazine, Talepipe.

500 (City of Winnipeg) Wing, Air Force Association of Canada Newsletter

http://www.wwnorton.com/college/history/america6_brief/dig_hist/airplanes/ Deposition by J. A. D. McCurdy, April 9, 1920, from the Alexander Graham Bell papers at the U.S. Library of Congress.

http://www.wwnorton.com/college/history/america6 brief/dig hist/airplanes/ Agreement to Organize the Aerial Experiment Association, from minutes by Thomas E. Selfridge, Oct.1, 1907.

Monsignor Donat Robichaud—De Moscou a Miscou, Revue de la Societe Historique Nicolas-Denys, vol XVII No 2, mai-aut 1989, pp47-74. Quarterly review of La Societe Historique Nicolas-Denys, a regional historical society of North-Eastern New Brunswick, mainly Gloucester County. Atlantic Canada Aviation Museum newsletter.

http://www.avroland.ca" \t "_blank" AVROLAND - A site dedicated to the people and aircraft of AVRO Canada & Orenda Engines Limited...

Maclean's Magazine

Errol Boyd, Canada's Lindy, by Herb Kugel, LOGBOOK, Aviation History, Vol. 4, No. 2, 2nd Quarter, 2003.

http://www.rareaviationphotos.com Biographical sketches and photos by Konrad Johannesson, RFC pilot, Olympic hockey player.

Pilot's Flying Log Book, Jim Shilliday

Index

No. 419 (Moose) Squadron, 85
3 Advanced Flying School, Gimli, 101
30 Air Materiel Base, Langar, Eng., 107

A

A.V. Roe Canada Ltd., 156, 157, 159
Advanced Flying School, Portage la Prairie, 113
Aerial Experiment Association 2005, 169
Aerial Experiment Association, 12, 20; disbanded, 22
"Aerodrome of Democracy", 83
Aeronautical Museum of the National Research Council, 23
Aerospace Museum, 169
Aileron, 22
Air Board Act (1919), 59
Air Board formed, 1919, 45
Air Board, 58, 63
Air mail contracts, 62
Air Transport Board, 63
Air-cooled engine, 22
Aircraft catapult, 15
Aircrew vets from five countries sue CBC, 92
Airforce Magazine, 12
Alcock, John, 67
Allen, Peter, 149
Allward, Walter Seymour, 30
Archibald's Hotel, 73
Armstrong-Whitworth FK-8 bomber, 42
Atherton, Flt. Lieut. Steve, 108
Atlantic Canada Aviation Museum, 77, 169
Atlantic mail and passenger service, 74-75
Aunt Georgina, 21
Austerity becomes prosperity, 100
Aviation and the Supreme Court, 94
Avro 504, 34
Avro Aircraft Ltd., 157
Avro CF-100, 151
Avro CF-105 Arrow, 12, 155, 158-59
Avro Lancaster sorties total, 90
Avro Lancaster, 81, 85, 90
Avro Manchester, 81

B

B-29 Superfortress, 118
Baddeck No. 1, 23
Baddeck, Nova Scotia, 12, 15, 22
Badger bomber, 119
Balbo Drive, 74, 75
Balbo, Gen. Italo, 74-75
Baldwin, Frederick Walker "Casey", 20-22
Barker, William, 26, 29, 31, 36, 47
Bauer, Harry, 101
Baumann, Alex, 151
Beagle bomber, 106, 109
Beinn Bhreagh, 15, 20
Bell, Alexander Graham (postage stamp), 56
Bell, Alexander Graham, 12, 15, 20, 90
Bell, Mabel, 20
Bellanca monoplane Columbus, 71
Biggles, 27
Birney, Earle, 15
Bishop, William "Billy", 31, 46
Black, Lt. Col. Dean C., 95
"Blind flying" course, 47
"Bloody April", 1917, 29
Boeing "Yankee Clippers", 76
Boeing Aircraft of Canada, 83
Boeing B-47 Stratojet, 118-119
Boeing Stearman, 90
Boll, Mabel, 71
Bombardier, 12, 161
Borden, Prime Minister R. L., 35
Botwood seaplane base, 76
Bourgeois, Joe, 88
Boyd, Erroll, 69-70
Boyd, Winnett, 156
Bras d'Or Lake, 12, 15
Breadner, Air Marshal L. S., 83
Bristol Aerospace, 149
Bristol Beaufighter, 81
Bristol Bolingbroke, 80, 82
Bristol Fighter, 43
British Commonwealth Air Training Plan, 59, 83, 100
Brock, William S., 69
Brounbrake, Pilot Officer M.K., 92
Brown, Arthur W., 67, 70
Brown, Roy, 31
Brown, Vernon, 35
Burgess-Dunne aircraft, 32

C

C-119 Flying Boxcar (436 Transport Squadron), 107
CAF No. 8 Aircraft Maintenance Squadron, 88
Camp Borden flight sheds, 34
Camp Borden, 27, 45, 47
Camp Petawawa demonstration, 23
Canada's centennial of flight year, 165
Canada's First Pilot, 15
Canadair CF-100 Canuck, 113
Canadair CL-215 and CL-415, 51
Canadair DC-4 North Star, 76
Canadair Sabre, 165
Canadair, 52
Canadian Aerodrome Company, 22
Canadian Aeroplane Company, 31
Canadian Aircraft Co., 57
Canadian Airways Ltd., 63
Canadian Associated Aircraft Ltd., 82
Canadian Aviation Corps, 31-32
Canadian Aviation Hall of Fame, 59
Canadian Aviation Historical Society, Manitoba branch, 12
Canadian Aviation Museum
Canadian Bushplane Heritage Centre, 169
Canadian Car and Foundry Co., 82
Canadian casualties with RAF squadrons, 92
Canadian Club, 22
Canadian Forces Air Command (AIRCOM), 147
Canadian National Railway, 55
Canadian Pacific Airlines, 60-61
Canadian Permanent Committee on Geographical Names, 93
Canadian Trans-Atlantic Air Service, 76
Canadian Vickers Ltd., 83
Canadian Warplane Heritage Museum, 85, 90
Canadians top RFC air aces, 38
Canberra bomber, 117, 119
Carrlon, Wally, 52
Carson, Nellie, 19
Catalina and Canson flying boats, 83
CBF Cold Lake, Alberta, 142
CC130 Hercules, 89
Centrifuge, 148, 149
CF-101 Voodoo, 147
CF-104 Starfighter, 147
CF-18 Hornet, 142, 145, 161
CFB Cold Lake, 148
CFB Trenton, 27
Chamberlin, James, 158
Chenier, E. J., 141
Chiasson, Father Ernest, 78
Chinook helicopter transport, 150, 155
Chretien, Prime Minister Jean, 146
Churchill, Manitoba, 57
Civil functions RCAF's primary role, 47
Clito, 143
Cobham, Sir Alan, 61
Cochrane, Jacqueline, 19
Cold War, 97, 102
Collins, Michael, 158
Comaircent, 119
Consolidated Liberator, 83
Cooke, Tom, 51
Cormorant helicopter, 150
Cotton, Sidney, 76
Coventry Civic Aerodrome, 153
Creelman, Flying Officer Ira "Slim", 116, 120
"Cross licensing", 22
Crossley, Earl, 51
Curtiss flying boats, 33
Curtiss HS-2L, 56
Curtiss JN-4 "Jennys", 28, 70; "Curtiss control", 22
Curtiss, Glenn W., 20-21, 56
Curtiss-Reid Courier, 63
Cutiss NC-TA flying boat, 67
Cygnet, tetrahedral kite, 21

D

Davis, Wing Commander Bob, 132
Day, Sen. Joseph, 92
De Havilland Aircraft of Canada, 54
De Havilland Beaver, 51, 60
De Havilland Mosquito, 81
De Havilland Otter, 52
De Havilland Tiger Moth, 82, 86
De Havilland Vampire, 115, 119
De Havilland Venom, 115
Defence Industries Ltd., 77
Dempsey, Dan, 168, 170
Dent, Major Len, 144
Department of National Defence, 147
Deseronto, 27
DEW Line, 100
Dickens, Punch, 31, 61
Diefenbaker, Prime Minister John, 146
Dilworth, Paul B., 156, 160
Douglas B-26, 50
Douglas Bader, 116
Douglas, W. A. B., 59, 81
Drake, Flying Officer Cal, 108, 119
"droming", 15
Ducarme, Morris, 88
Dunkirk evacuation, 92
Durrell, Lawrence (Bitter Lemons), 143

E

Earhart, Amelia, 19, 65-66, 71, 73
Elliot Brothers, 57
Enders, Capt. George, 72
Eskimo Point, 53
Exercise Carte Blanche, 137
Exercise Dividend, 118, 121
Expeditor, 110
Eyebrook Reservoir, 81

F

F-100 Super Sabre, 134

F-86 Sabre squadrons, 30, 105; serving in Europe between 1951 and 1963, 107
Fairchild Aerial Surveys Co., 56
Fairchild Aircraft of Canada Ltd., 82
Fairchild Company, 54
Fairchild Super 71, 54
Farnborough Air Show, 129
Farr, Harry, 32
First aircraft plant in U.S., 24
Fisher, Squadron Leader Art, 135
Fitz Gerald, Patrick, 55
Fleming, Flying Officer Bruce, 108, 139
Floyd, Jim 158
Flying "blind", 84
Flying clubs, 59; organize for war, 86
Fokker Universal, 57
Franks, Dr. Wilbur, 149
French Mystere, 134
French-Canadians, 91

Gander Airport, 76
Gatineau airfield, 166
Gatty, Harold, 72
Gill, George, 51
Gilruth, Robert, 158
Gloster Meteor, 115
Gordienko, Maj. Mikhael, 78
Graf Zeppelin, 72
Granatstein, J. L., 12
Grayson, Frances, 71
Green, H. Gordon, 15
Grosvenor, Gilbert, 24
Grumman Avenger, 49
Grumman Goose, 79
G-suits, 149
Guillet, G. R., 100
Guynemer Trophy, 133
Gypsy Moth, 54

H
Haddon, Gerald, 169
Hadfield, Col. (Ret.) Chris, 168, 170
Halliday, Hugh A., 59, 138
Hammond, Lt. Arthur, 42
Hammondsport, N.Y., 15, 22
Handley Page Atlantic, 67
Handley Page Halifax NA 337, 86, 87; Raising NA 337, 88
Handley Page Halifax, 81
Handley Page Hampden, 81-82
Harbour Grace Airport Trust Company, 68
Harbour Grace, 64-65, 71
Harmon Trophy (1927), 57
Harper, Prime Minister Stephen, 146, 151
Harvard Mark 4, 86
Hawk One, 165
Hawker Hunter, 118, 134
Hawker Hurricane, 82, 90

Hawn, Lawrie, 149
Haycock, Ronald, 35
Hayes, Brig.-Gen. (Ret.) Paul, 170
Heavy lift C-17, 150
Hellyer, Defence Minister Paul, 146
Henry, Ivan, 97
Hepburn, Mitchell, 62
Hercules C-130J, 150
Hindenburg, 76
Hornell, Flt. Lt. David, 83
Horsa and Hamilcar gliders, 81
Howe, C. D., 63
HS-2L flying boats, 58
Hudson Bay Railway, 57
Hughes, Sam, 31-32, 35, 55
Humber-Bleriot XI, 153

I
Imperial Airways, 76
Imperial gift of aircraft to Canada: Avro 504; Royal Aircraft Factory SE-5A; de Havilland DH-4 and DH-9; Bristol 5-2B; Sopwith Camel; Felixstowe F-3; Curtiss H-16 flying boat; Fairey C-3; Sopwith Snipe, 44
International Civil Aviation Association, 59

J
James Armstrong Richardson International Airport, 57
Janney, Ernie, 32
Jeffrey, Jeff, 88
Johannesson, Konnie, 37, 94
Johnson, Amy, 19, 73
Johnson, Bill, 116
Johnson, Vic, 12, 142
Johnston, R. N., 51
June Bug, 21,
Junkers, 53

K
Kelly, Flt. Lt. Dean, 49, 154
Kerr, Rear Admiral Sir Mark, 67
King Edward, 19
King, Prime Minister W. L. Mackenzie, 62, 71
Kirschstein, Lt. Hans, 42
Kissmann, Paul, 168, 170
Kittyhawk, N.C., 15
Kjarsgaard, Karl, 88
Koehler, Fred, 68-69, 71
Kokkinaki, Brig.-Gen. Vladimir Konstantinovich, 77-78, 106
Korean war, 102, 146
Kozak, Flying Officer Fred, 120

L
Lac du Bonnet, Manitoba, 52, 61
Lac-a-la-Tortue, Quebec, 56
Lake Keuka, 22

Lancaster XPP, 76
Langley, Samuel, 22
Langruth firing range, 104
Laurentide Air Service, 56
Leavens Aviation Inc., 169
Leckie, Col. Robert, 45
Leslie, Tim, 164, 168, 170
Levesque, Flight Lieutenant Omer, 138
Lindbergh, Anne, 19
Lindbergh, Charles, 62, 68
Lindley-French, Dr. Julian, 150
List of world's first pilots: Wright, Orville; Santos-Dumont, Alberto; Bleriot, Louis; Farman, Henri; Baldwin, F.W. (Casey); Delagrange, Leon; Selfridge, Lt. Thomas; Curtiss, Glenn; McCurdy, J.A. Douglas; Roe, Alliott Verdon, 24
Litvyak, Lilya, 19
Lockheed (Canadair) CF-104 Starfighter, 90
Lockheed 14, 79
Lockheed Electra, 63
Lockheed T-33 trainer, 27, 97
Lockheed Vega, 66, 68, 72
Lockheed-Martin F-35 Joint Strike Force fighter, 149, 150

M
MacBrien, Gen. J.H., 47
MacDermid, John, 15
Magee, John Gillespie, 86
Magill, Eileen, 19
Magyar, Lt. Alexander, 72
Map, West and East Germany: No. 1 Fighter Wing, Marville, France; No. 2 Fighter Wing, Grostenquin, France; No. 3 Fighter Wing, Zweibruken, W. Germany; No. 4 Fighter Wing, Baden Soellingen, W. Germany.
Marion, Normand, 34
Markham, Beryl, 19, 73
Mars Rover, 160
Marsh, James H., 157
Marsoe, Tore, 88
Martin, Prime Minister Paul, 146
Marville, France, 30
May, Wop, 31, 58
Maynard, Owen, 158
McCallum, Don, 116
McCurdy Flying School, 33
McCurdy, J. A. Douglas, 12, 15, 20, 22, 24, 31, 36, 55-56, 63, 70, 82, 169
McDairmid, J. G., 62
McDonnell-Douglas CF-18 Hornet, 147
McElmon, Flying Officer Douglas "Moose", 121
McKay, Flying Officer Ev, 116

McKee trophy, 24
McKey, Alys, 19
McKnight, William Lidstone, 91
McLeod, Alan, 28, 31, 40,
McLeod, Bruce, 116
Me-109, 92
Me-110, 92
He-111, 92
Meighen, Sen. Michael, 92
Mepham, Flying Officer Pat, 108, 119
Merging of Canadian forces, 146
Mersereau, Cyril, 79
Metcalfe, Lt. James, 69
Mid-air collision, 127
Mid-Canada Line, 100
MiG 15, 103
Mig-17, 103
MiGs versus Sabres, 138
Milberry, Larry, 12, 133
Miss Dorothy, Bellanca monoplane, 69
Mockford, Dennis, 170
Mockford, Greg, 170
Mollison, Jim, 69, 73
Molson, K. M., 31, 55
Morgan, Bob, 116
Morton, Desmond, 106, 147
Moscow to Miscou flight, 76
Moskvka, Illyushin TsKB bomber, 77
Mussolini, Benito, 74
Myasishchev Bison bomber, 118
Mynarski, Pilot Officer Andrew, 85, 90

N
NASA space program, moon landings, 12
National Steel Car Company, 82
NATO, 105, 115, 145-146
New York World's Fair, 77
No. 1 (Fighter) Wing, RCAF, 81
No. 1 Air Division H.Q., Metz, France, 107
No. 1 Fighter Wing, Marville, France, 136
No. 1 Fighter Wing, North Luffenham, 107, 153
No. 1 Flying Training School, Centralia, 100, 110
No. 1 Pilot Weapons School, Macdonald, 113
No. 10 Squadron (North Atlantic), 83
No. 2 Fighter Wing, Grostenquin, France, 107, 136
No. 242 Squadron, RAF, 91
No. 3 Wing, Zweibruken, 127
No. 4 Fighter Wing, Baden-Soellingen, W. Germany, 105, 130, 145
No. 410 (Cougar) Squadron, 48, 81, 114, 121, 126, 130,
No. 435 Transport Squadron, 108

No. 439 (Sabre Tooth Tigers) Fighter Squadron, 114
No. 439 Squadron, 81
No. 441 (Silver Fox) Fighter Squadron, 81, 114
No. 441 Squadron Sabres, 132
No. 644 Squadron RAF, 87
Noorduyn Norseman, 48, 52, 54
NORAD, 150
North American Harvard, 90, 97, 110
North Luffenham, 81, 84
Northcliffe, Lord, 67
"Nose hangar", 55

O
Oke, John L., 68
Oldfin, Flt. Lt. Ray, 99,108
Ontario Trillium Foundation, 169
Operation Vanity across Rhine River, 87
Orenda (10, 14, Iroquois), 155
Orenda Engines Ltd., 157
Orenda Iroquois engine, 152
Ottawa Car Company 77
Ottawa Car Manufacturing Co., 82

P
Pan-American Airlines, 76
PANAVIA Tornado, 147
Patent rights to McCurdy and Baldwin, 22
Pearson, Prime Minister Lester, 146
Peden, D. Murray, 84
Portal, Charles, chief of air staff, 35
Post, Wiley, 68, 72
Potter, Michael, 165, 166, 170
Pratt and Whitney Canada, 169
Preparing for war, 105
Princess Patricia Canadian Light Infantry, 143

R
RAF Bomber Command, 84
RAF Coltishall, 104, 119
RAF Gloster Meteors, 104
RAF Station Scampton, 93
RAF Station Tarrant Rushton, 86
RAF Station Worksop, 93
RAF Station West Raynham, 118
RAF Station Wittering, 127
Raid on Zeitz, Germany, 93
Ralph, Wayne, 36, 171
RCAF Boeing 707, 143
RCAF Memorial Museum, 88
RCAF Station Gimli, 28
RCAF's No.1 Air Division, 12

Read, Lt.-Cdr. Albert, 67
Red Lake gold district, 58
Red Wing, 21-22
Reichers, Lou, 72
Reid Aircraft Co., 54
Resistance forces, "packages" and "Joes", 87
Reynolds Alberta Museum, 169
RFC Canada, 34
RFC/RAF, 31
Richardson, James A., 55, 58, 62
Richenbacker, Eddie, 73
Robichaud, Antonine, 78
Robichaud, Monsignor Donat, 77
Rogers, Will, 72
Roosevelt, Pres. Franklin Delano, 83
Roosevelt, Pres. Theodore, 21
Rosevear, Capt. Stanley Wallace, 42
Royal Canadian Air Force formed, 1924, 47
Royal Canadian Legion, 95
Royal Flying Corps, 28-29, 34
Royal Naval Air Service, 33
Russia, 103
Rutley, Capt. Doug, 88

S
Sabre fatalities, 138
Sabre Pilots Association of Air Division Squadrons (SPAADS), 170
San Antonio Gold Mine, 61
Savoia-Marchetti SM-55X flying boat, 74
Schiller, Duke, 69
Schlee, Edward, 69
Secret Operations Executive, 87
Segal, Sen. Hugh, 92
Selfridge, Lt. Thomas E., 20-22, 24
Separate Canadian air force rejected, 35
Shannon, Norman, 32
Sharpe, William, 32
Shilliday Lake, 93
Shilliday, Flight Sergeant Robert Charles, 93
Shilliday, Jim, 116
Sikorsky S-42, 76
Silver Dart, 15, 21-22; replica, 90; replica flies, 169
Simpson, Jeffrey, 161
Sir Frank Whittle Jet Heritage Centre, 155
Slessor, Air Chief Marshal Sir John, 35, 102
Smith, Ben, 62
Smye, Fred, 156, 157

Smyth, Ross, 71
Sopwith 1 l/2 Strutter, 76
Sopwith Camel, 26, 29
Sopwith Pup, 90
SPAADS, 99
Space robotics, 160
Spar Aerospace "Canadarm", 160
Speke airfield, 87
Spirit of Harbour Grace, 73
Spirit of St. Louis, 68
St. Laurent, Prime Minister Louis, 146
Stalin, Joseph, 76
Steen, John, 88
Stevenson Field, 57, 94, 155, 156
Stevenson, Frederick J., 31, 57
Stinson Aircraft Corporation, 68
Stinson Reliant, 51
Strategic Aerospace and Defence Initiative, 149
Supermarine Spitfire, 83
Sweetman, Flying Officer Don, 97

T
Talbot, J. H., first flying fatality, 35
Tarling, "Turbo", 49
Taylor, H. A., 55
Tegami base, Ontario Provincial Air Service, 51
Tegart, Flying Officer Al, 120
The Valour and the Horror, second episode—*Death by Moonlight: Bomber Command,* 92
Thibault, Brig.-Gen. Claud, 147
Thorne, Wilf, 116, 127
Three Bs—Bishop, Barker, Brown, 38
Toews, Vic, 57
Trans-Canada Airlines, 63, 76
Trudeau, Prime Minister Pierre, 146
Tryggvason, Bjarni, 169
Tully, Capt. Gerry, 69
Tupolev Bear bomber, 118
Turnbull, Wallace, 47
Tytula, Lt. Col. Bill, 89

U
U-2 spyplane, 141
UN Forces in Cyprus, 143
Unmanned Aerial Vehicles (UAV), 150
USAF Base Sculthorpe, 118

V
Vaesen, Flying Officer "Tweedie", 108
Vance, Johnathan F., 86, *High Flight,* 12
Vanderbilt, Harold S., 79
Variable-pitch propeller, 47

Venus, Flying Officer Bud, 120
Vibert, Laurence, 79
Vickers Gunbus, 35
Vickers Vedette, 54
Vickers Vimy, 67
Vickers Wellington, 81
Victoria Beach air force base, 45
Victory Aircraft Company, 76
Villeneuve, Flt. Lt. Fern, 123, 168
Vimy Monument, 30
Vimy Ridge, 29
Vintage Wings of Canada aircraft stable, 167
Vintage Wings of Canada Foundation, 165
Vollick, Eileen, 19
Von Richthofen, Baron Manfred, 31
Vulcan bomber, 119

W
Walmsley, Air Marshal Sir Hugh, 35
Wartime Measures Act, 77
Wavell, Field Marshal Lord, 146
Western Canada Airways, 53, 57-58, 60-61, 63
Western Canada Aviation Museum, 54
Western Front artillery bombardment, 29
Westland Lysander, 82
Westphal, Flying Officer Jerry, 133
White Wing, 21
Will, Lt. Col. Steve, 168
Williams, Bernard, 112
Williams, James N., 86
Wilson, Ellwood, 56
Wilson, J. S., 47
Wilson, John A., 59, 94
"wing flapping" 15
"wing-warping", 22
Winnipeg Falcons, 94
Winnipeg Flying Club, 94
Wise, S. F., 34
Women flyers, 19
Wood, Philip, 69
Wright Brothers 1903 flight, 12
Wright Flyer, 15
Wright, Lloyd, 88

Y
Yeager, Chuck, 115

Z
Zulu alert, scrambles, 108 139